GLOSSARY OF ACADEMIC VOCABULARY FOR IB DIPLOMA

Maths: Applications and Interpretation

JANE APPLETON
LEISA BOVEY
SPYRIDON KITSIONAS
MARGARITA SIFNAIOU

Published by Elemi International Schools Publisher Ltd

© Copyright 2019 Elemi International Schools Publisher Ltd

First published 2019

Authors: Jane Appleton, Leisa Bovey, Spyridon Kitsionas, Margarita Sifnaiou
Series Editor: Mary James

The authors and publisher would like to thank the following people for their valuable contributions: Amanda George who is an experienced tutor and writer of physics and maths educational content for textbooks, digital platforms and exam papers; and Tilly Mitchison, a former IB Diploma Maths student.

We are a small independent publishing company. If you photocopy this book or if you are given a photocopied version of this book or any part of it, please be aware that you are denying the authors and the publisher the right to be appropriately paid for their work. However this book found its way to you, we'd welcome any feedback you'd like to share with us: please use the contact form on our website.

All rights reserved. No part of this publication may be copied, reproduced, duplicated, stored in a retrieval system, or transmitted in any form or by any means, without the prior written permission of Elemi International Schools Publisher Ltd, or as permitted by law or by licence. Enquiries about permission for reproduction should be addressed to the publisher.

A catalogue record of this title is available from the British Library
British Library Cataloguing in Publication Data

ISBN 978-1-9164131-3-9

10 9 8 7 6 5 4 3 2 1

Page layout/design by emc design ltd.
Cover design by Jayne Martin-Kaye

Printed and bound in Great Britain by TJ International Ltd, Padstow, Cornwall

We are an entirely independent publishing company. This resource has been developed independently from and is not endorsed by the International Baccalaureate Organization. International Baccalaureate, Baccalauréat International, Bachillerato Internacional and IB are registered trademarks owned by the International Baccalaureate Organization.

Studying Mathematics: Applications and Interpretation at IB Diploma level

The IB Diploma programme Mathematics: Applications and Interpretation is a rigorous and challenging course of study which helps build your mathematical knowledge and understanding as well as your problem-solving skills. A key part of your studies will involve developing communication, interpretation, and reasoning skills through mathematical inquiry and argument and will require you to use appropriate mathematical terminology in precise statements.

How this resource can help you

Studying Mathematics: Applications and Interpretation as part of the IB Diploma programme involves a substantial amount of time for independent study and you may need additional support from your teacher, friends, or other resources. Of course, your teacher and friends may not always be available, particularly when it comes to acquiring, learning, and using the mathematical language and terminology from the course.

This book aims to help you in this process by unpacking the language of the IB Diploma with a focus on both course content and assessment.

- Each word or phrase included in this A-Z glossary has been carefully selected because we think it will be useful in your studies. This resource contains words and phrases from the IB Diploma Mathematics: Applications and Interpretation subject guide as well as subject-specific terms commonly found in most Maths textbooks.

- You will find key subject-specific vocabulary related to your study of Maths, such as mathematical terms, concepts, and techniques, as well as command terms and assessment terminology. Knowledge and appropriate use of these terms will have a significant impact on your overall achievement and final score.

- Terms related to the IB Learner Profile are also included and explained within the context of your Maths studies, so that you may understand how these learner attributes are applied within the Maths classroom.

- Support is offered on using your graphical display calculator with many of the abbreviations used on your calculator explained within this glossary, too.

- Your IB Diploma Maths course is divided up into five topics: Number and algebra, Functions, Geometry and trigonometry, Statistics and probability, and Calculus. To help your understanding of Maths, the relevant topic is given in brackets at the end of each definition. If no topic is provided, this is because you will find yourself using that term across a number of different topics in Maths.

- All of the terms in this resource will be appropriate for students studying at Higher Level. Words and phrases that are labelled (AHL) indicate content that is additional HL; this means that HL students will come across this language in their studies, but these terms do *not* form part of the SL course. Students studying at SL need only focus on those terms which are *not* highlighted (AHL).

- Where you see a word in the definition written in **green text**, this means that a glossed definition exists for it elsewhere in the book. This has been done where we thought it would be helpful for you.

- Do remember that this resource is *not* a dictionary, as it does not necessarily contain all possible definitions for each word or phrase. It is, however, a glossary of terms where the definitions are given in the context of the IB Diploma Mathematics: Applications and Interpretation course.

- Also note this is *not* a comprehensive list of mathematical terms. If your teacher gives you some additional words, you might choose to write them into the glossary yourself, so that the book is more like a living workbook for you.

- Your teacher might also encourage you to extend the current list with additional terms, enhance the definitions according to their own ideas and interpretations, or provide alternative examples.

We wish you the best on your learning journey and of course the greatest success in your exams!

Jane, Leisa, Spyridon, Margarita, and the team at Elemi

1D	One dimension (1D) is when we measure in one direction, usually length (a line). In calculus, used to describe an object that moves only in one direction.
2D	Two-dimensional. Shapes like squares, rectangles, etc are two dimensional as they are flat shapes. The shapes have two dimensions (eg height and width, but no depth). (Geometry and trigonometry)
3D	Three dimensional. Shapes like cubes, cuboids, spheres, etc are three dimensional as they have three dimensions (eg height, width and depth). (Geometry and trigonometry)
Absolute extremum (*plural* absolute extrema)	The largest or smallest **value** that a **function** may take. (Calculus)
Absolute value	The size of a number regardless of whether it is positive or negative. The absolute value of a number will always be positive. For example, the absolute value of −30 is 30. (Also known as **modulus**.) (Number and algebra)
Absorbing Markov chain (AHL)	A **Markov chain** in which there is at least one **absorbing state** and it is possible from every state to reach an absorbing state. (Statistics and probability)
Absorbing state (AHL)	A state in a **Markov chain** that cannot be left once entered. For example, state C in the Markov chain shown below is an absorbing state because once you reach state C, you cannot get out of it. (see **periodic state**)

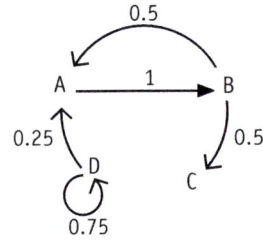

(Statistics and probability)

Abstract (abstraction)	To take away any element of real life to which a mathematical concept might be related and then generalize it so that it can be more broadly applied. (Number and algebra)
Acceleration (AHL)	The rate of change of the **velocity** of an object per unit of time. (Geometry and trigonometry, Calculus)
Acceleration function (AHL)	A function that describes the **acceleration** of an object at time t. The acceleration function is the **second derivative** with respect to time of the **displacement function**. If $s(t)$ is the displacement function, then $v(t) = \frac{ds}{dt}$ is the **velocity function** and the acceleration function is $a(t) = \frac{dv}{dt} = \frac{d^2s}{dt^2}$. (Calculus)
Acceptance region	The values which make the **null hypothesis** valid, and is expressed as an upper and lower limit. (Statistics and probability)
Accurate (accuracy)	How close a measured value is to the actual (true) value. An exam question will tell you how the answer should be given – this is the **degree of accuracy**. If a question asks for an exact answer, the answer should be given in full or as a **surd** if necessary. (Number and algebra)
Acute angle	An angle which measures between 0 and 90 degrees. (Geometry and trigonometry)

Acyclic (AHL)	A graph that does not have a **cycle**, ie it does not have a **path** that begins and ends in the same place without revisiting any **vertices** (other than the start and end). (Geometry and trigonometry)
Addition (add)	Finding the sum of numbers. For example: 3 add 4 is $3 + 4 = 7$. (Number and algebra)
Additive identity (AHL)	The property stating that the **sum** of any real number and 0 is equal to the original number: $a + 0 = 0 + a = a$, for all real numbers. In this case, 0 is called the additive identity element. The additive identity property is extended to other sets as well, such as complex numbers, where the identity element is also 0, and matrices, where the identity element is the **zero matrix** O. (Number and algebra)
Additive inverse (AHL)	Every number has its **opposite** as an additive inverse. The sum of a number and its additive inverse is equal to the additive identity element, ie $a + (-a) = 0$. Similarly, every matrix A has matrix $-A$ as its additive inverse, and their sum is equal to the **zero matrix** O. (Number and algebra)
Address (AHL)	The **ordered pair** (i, j) stating the position of an **element** a_{ij} of a matrix. This lies in row i and column j. For example, the address of the element -6 in the matrix $\begin{pmatrix} 3 & -4 \\ -6 & 5 \end{pmatrix}$ is $(2, 1)$. (Number and algebra)
Adjacency matrix (AHL)	A matrix used to represent a finite graph. It shows the connections between **vertices** and whether or not the vertices are **adjacent vertices**. For a graph with V vertices, the matrix is a $V \times V$ matrix. An entry of 1 in the (i, j) entry shows that the vertices i and j are connected by an edge and an entry of 0 shows they are not. For example:

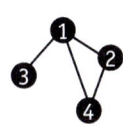

$$\begin{array}{c} \\ 1 \\ 2 \\ 3 \\ 4 \end{array} \begin{array}{c} 1 \quad 2 \quad 3 \quad 4 \\ \begin{bmatrix} 0 & 1 & 1 & 1 \\ 1 & 0 & 0 & 1 \\ 1 & 0 & 0 & 0 \\ 1 & 1 & 0 & 0 \end{bmatrix} \end{array}$$

Powers of an adjacency matrix can be used to find the number of k-step walks between two vertices. For example, given A as the adjacency matrix, A^4 would give the number of 4-step walks. (Geometry and trigonometry)

Adjacent	This commonly means 'next to' but in trigonometry, this refers to the side of a right triangle which is neither the **hypotenuse** nor the side opposite the angle we are using in the sum. (Geometry and trigonometry)
Adjacent edges (AHL)	In **graph theory**, adjacent edges are edges that share a common **vertex**. (Geometry and trigonometry)
Adjacent vertices (AHL)	**Vertices** that are connected to each other by an edge. (Geometry and trigonometry)
Algebra	The use of letters to replace numbers in order to work out a sum. For example, to work out how many sweets there are in a certain number of bags, we might say that if each bag contains c sweets, then a bags would contain $a \times c$ or ac sweets. (Number and algebra)
Algebraic expression	A mathematical expression consisting of numbers, letters and **operations**. For example: $5a + 2c$ (Number and algebra)

A

Algebraic fraction (AHL)	A fraction where the numerator and denominator are **algebraic expressions**. For example: $$\frac{2x}{x+1} \qquad \frac{a+3b}{6a-b}$$ (Number and algebra)
Algorithm (AHL)	A set of steps to be followed in order to solve a problem. (Geometry and trigonometry)
Alternative hypothesis	The hypothesis you accept if a **hypothesis test** causes you to reject the **null hypothesis**. It is the opposite of the null hypothesis. For example, if your null hypothesis (H_0) is that cereal choice and gender are independent, the alternative hypothesis (H_1) would be that cereal choice is dependent on gender. (Statistics and probability)
Ambiguous case (of sine rule) (AHL)	When using the sine rule results in two possible angles, which would create different triangles. This happens when you are given two sides and an angle that is not between the two sides. (Geometry and trigonometry)
Amortization	Used in finance eg in loans repayment, when the value is spread out or decreases over periods of time. (Number and algebra)
Amplitude	Half the distance between the lowest and highest values of a function. For example, in the graph of sine x, which goes from -1 to $+1$, the amplitude is 1. (Functions)
Analyse (analysis)	To look at something in detail to try and explain it in mathematical language so that you are able to work out a solution.
Angle	Formed by two lines where they join at a point. The size of angles are measured in **degrees** or **radians**. (Geometry and trigonometry) (see **vertex**)
Angle of depression	When you are looking at something below your eye level you are looking down at an angle of depression. (Geometry and trigonometry)
Angle of elevation	When you are looking at something above your eye level you are looking up at an angle of elevation. (Geometry and trigonometry)
Annual	Something that happens once every year or that relates to the period of a year. (Number and algebra)
Annual depreciation	An amount (usually a percentage) by which an amount decreases each year. (Number and algebra)
Annual inflation	An amount (usually a percentage) by which an amount increases each year. (Number and algebra)
Annuity (annuities)	An amount of money paid to you, often over your life, maybe in monthly instalments. You can buy a lifetime annuity, for example, which will pay you monthly. (Number and algebra)
Annuity formula	The formula used to calculate the amount of money to be paid in instalments on an **annuity**. (Number and algebra)
Answers matrix (AHL)	The matrix that is the answer to a matrix equation of the form $AX = B$. (Number and algebra) (see **coefficient matrix**, **constant matrix**, **variable matrix**)
Antiderivative	The opposite of **derivative**; this is what you get when you **integrate**. (Calculus) (see **integration**)
Antidifferentiation	The process of finding the **antiderivative**. (Calculus)

Not to be photocopied

Aperiodic state (AHL)	A state in a Markov chain that is not a **periodic state**. For an example, see **periodic state**. (Statistics and probability)
Application	How a mathematical process can be applied to 'real life'. For example, when estimating the height of a tree, we can measure the distance to the tree in metres, use a clinometer and work it out using **trigonometry**.
Appropriate (data) (AHL)	Data that is suitable for the intended purpose of analysis. (Statistics and probability)
Approximate (approximation)	An estimation, not exact. (Number and algebra)
Arabic system (of numbers)	The numbers 0, 1, 2, 3, 4, 5, 6, 7, 8, 9 and numbers formed from them. (Number and algebra)
Arc	A part of the circumference of a circle, or a **segment** of any curve. (Geometry and trigonometry)
Area	The amount of space inside a shape. Square units are used to measure area, for example, cm², m². (Geometry and trigonometry)
Argand diagram (AHL)	The geometrical representation of a complex number $z = x + iy$ as a point with coordinates (x, y) or a position vector $\begin{pmatrix} x \\ y \end{pmatrix}$ on the Argand or **complex plane**. (Number and algebra)
Argument (of a complex number) (AHL)	A complex number $z = x + iy$ can be written as: $$z = r(\cos\theta + i\sin\theta) = re^{i\theta}.$$ It has argument θ, where $\arg z = \theta = \tan^{-1}\left(\frac{y}{x}\right)$, ie the counter-clockwise angle formed between the **horizontal** axis and the **vector** $\begin{pmatrix} x \\ y \end{pmatrix}$ representing the complex number on the Argand diagram or **complex plane**. The angle θ is measured in **radians**. (Number and algebra)
Argument (of a function)	A term that you put into a function and which will affect the result of the function. For example, for $\sqrt{x}$, the argument is x. (Functions)
Arithmetic	The use of numbers in Mathematics for addition, subtraction, division, etc. (Number and algebra)
Arithmetic progression	Another name for **arithmetic sequence**. (Number and algebra)
Arithmetic sequence	A sequence of numbers which are formed by adding the same number (or **common difference**) each time. For example, in the sequence 1, 5, 9, 13, 17, the common difference is 4. The common difference can be positive or negative. Also called arithmetic progression. (Number and algebra)
Arithmetic series	The sum of the numbers in an **arithmetic sequence**. (Number and algebra)
Array (AHL)	A data structure consisting of objects of the same type (eg values or variables). An array is used to store a collection of data. A two-dimensional array is called a **matrix**. A one-dimensional array is called a **vector**. (Number and algebra)
Ascending order	In order from the lowest number to the highest. (Number and algebra) (see **descending order**)

A

Associative (associativity) (AHL) — The property stating that an **operation** on three or more objects is independent of how the objects are grouped. Both addition and multiplication of **real** or **complex numbers** are associative:

$$a + (b + c) = (a + b) + c$$

$$a \times (b \times c) = (a \times b) \times c$$

However, subtraction and division are not!

The associative property can be similarly applied to other sets, such as **matrices** and **vectors**. For example (for matrices of appropriate dimensions),

$$\boldsymbol{A} + (\boldsymbol{B} + \boldsymbol{C}) = (\boldsymbol{A} + \boldsymbol{B}) + \boldsymbol{C}$$

and

$$\boldsymbol{A} \times (\boldsymbol{B} \times \boldsymbol{C}) = (\boldsymbol{A} \times \boldsymbol{B}) \times \boldsymbol{C}$$

(Number and algebra, Geometry and trigonometry)

Asymmetrical (asymmetry) — The opposite of **symmetrical** (**symmetry**). An asymmetrical shape does not divide into two identical halves – it has no mirror line.

Asymptote — A line which a **function** approaches and although it nearly touches, it never quite does. (Functions)

Average — There are many types of averages but the most common one is the **mean** – you add together a series of numbers and then divide by how many numbers there are. Other averages commonly used are **median** and **mode**. (Statistics and probability)

Axiom (AHL) — A statement that is taken to be true without the need for proof or justification. It is considered to be a fact or an assumption that is used as a starting point to build further knowledge and mathematical **theorems**. An example of an axiom that you will use in your studies is the additive axiom, which states that if $a = b$ and $c = d$ then $a + c = b + d$. Axioms are also known as postulates, assertions or premises.

Axiomatic system (AHL) — A system of axioms on which a field of Mathematics can be based or developed. The most famous example is Euclid's five postulates, on which **Euclidean geometry** is based. These are: 1) there is a straight line connecting any pair of points; 2) each line segment can be extended to infinity in either direction; 3) a circle is defined by its centre and radius; 4) all right angles are equal in size; 5) the parallel postulate (ie if two lines are intersected by a third line, on one side of which the sum of the angles formed with the original lines is not exactly equal to two right angles, then the original two lines will eventually intersect each other).

Axis (plural axes) — A line on which a graph is based, and which enables you to describe the position of a point. For **2D** graphs, we use the x-**axis** and the y-**axis.** (Geometry and trigonometry)

Axis of symmetry — A line which cuts a shape into two identical halves. (Functions)

Balanced	Making sure that you work and 'play' in the right ratio. One of the qualities from the IB learner profile. In Mathematics, you need to make sure that you work for a certain amount of time and then take a break so your brain can recharge, ready to tackle the more difficult questions. In addition, when you are working on your mathematical exploration, you will need to have a balanced argument.
Bar chart (or bar graph)	A simple type of graph where rectangular bars are used to represent numbers in a category visually. The bars should usually have gaps between them. (Statistics and probability)
Base	When working with indices, the base is the number that is raised to a power. For example, for 2 cubed (2^3), the base number is 2. (Number and algebra)
Base unit	The international system of units used for mass, length, etc – often referred to as SI units. (Number and algebra)
Base vector (AHL)	A vector with a magnitude of 1 in the x- y- or z-direction. In the x-direction, the base vector is i, in the y-direction it is j and in the z-direction it is k. (Geometry and trigonometry)
Bearing	An angle which has three digits and is measured clockwise from the North line. So: North is 000°
	East is 090°
	South west is 225°
	(Geometry and trigonometry)
Best fit line (or line of best fit)	A straight line drawn on a scatter diagram to show a relationship between points. The line should go through as many of the points as possible. (Statistics and probability)
Best-fit straight line (AHL)	For bivariate data that has been produced as a result of logarithmic scaling on variables with values that extend over several orders of magnitude, one can apply linear regression, and through linearization reveal a power or an exponential relationship between the original variables (ie those before scaling), for a log-log and a semi-log graph, respectively. (Functions)
Bias	When the outcomes of an experiment are not reliable because the experiment is not fair or has a built-in error. For example, standing outside a supermarket and asking those who come out what their favourite supermarket is would lead to bias, as those who have shopped there are more likely to name it. (Statistics and probability)
Biased	You have a biased die where the chance of scoring each number from 1 to 6 is not equal. (Statistics and probability)
	(see bias)
Biased estimate (AHL)	When the value of a statistic from a sample does not equal the true value of that statistic from the population. For example, the variance of a sample is a biased estimate of the population variance, that's why there is a formula to estimate the population variance, σ^2, given the sample variance, s^2 (this formula is given in your formula book). (Statistics and probability)
	(see unbiased estimate)
Biased estimator (AHL)	A statistic from a sample that gives a biased estimate. (Statistics and probability)
Biased sample	A sample which has not been gathered fairly or has error built in. For example, standing outside a particular supermarket and asking people to name their favourite supermarket is likely to provide you with a biased sample. (Statistics and probability)
	(see bias)

B

Billion	A thousand million, 1,000,000,000 or 10^9. (Number and algebra)
Billionth	A thousand millionth, 1/1,000,000,000 or 0.000,000,001 or 10^{-9}. (Number and algebra)
Binomial distribution	A diagram to represent binomial probability – where there are only two possible outcomes, for example, tossing a coin. (Statistics and probability)
Binomial probability	The probability of getting a specific number of successes when you do a fixed number of trials. For example, the probability of getting 4 heads when you toss a coin 10 times. (Statistics and probability)
Binomial test (AHL)	A test that uses the probabilities from a **binomial distribution** to determine how significantly data from an experiment differs from expected outcomes. The binomial test is useful when there are only two possible outcomes. You need: • the number of trials, n • the number of successes, k • the probability of success, p • and the desired significance level. Use the **binomial distribution** to find the probability of the experimental outcome happening by **chance** and compare it to the desired **significance level**. (Statistics and probability)
Bipartite graph (AHL)	A graph of which the **vertices** can be split into two subsets where each edge from one subset only connects to a vertex from the other subset. (Geometry and trigonometry)
Bisect	To cut in half. (Geometry and trigonometry)
Bisector	A line which cuts a line, angle, or shape in half. (Geometry and trigonometry)
Bivariate data	Pairs of data involving two **variables** (for example, height and weight) which may be linked. (Statistics and probability)
Bivariate normal distribution (AHL)	A **normal distribution** with two independent random variables, both of which are normally distributed. (Statistics and probability)
Bivariate statistics	Data with two variables (for example, height and weight). Also called **bivariate data**. (Statistics and probability)
Bound	The ends of intervals in which an answer can lie – usually called **upper bound** and **lower bound**. (Number and algebra)
Boundary (on Voronoi diagram)	The lines between the regions of a **Voronoi diagram**. (Geometry and trigonometry)
Boundary condition	A limiting value used to find a solution. (Calculus)
Bounded	Surrounded on all sides. (Geometry and trigonometry)
Box and whisker diagram (*or* Box plot)	A diagram in statistics to represent a set of data using the minimum and maximum values, the lower quartile, the median, and the upper quartile. (Statistics and probability)
Brackets	These are brackets: ( ). They usually mean that you have to work out the sum inside them first. (Number and algebra) Brackets can also be used in Mathematics to contain coordinates, for example (5, 3). (Geometry and trigonometry)
Buying rate	The price for which something is purchased. (Number and algebra)

Calculate	Work out, not necessarily with a calculator. For example, if asked to calculate the sum of 2 and 3, this would mean that you add 2 and 3 to make 5. Can be used as a **command term** at IB Diploma.
Calculus	A part of Mathematics which looks at **differentiation** and **integration**. (Calculus)
Caring	The IB learner profile suggests ten qualities that students should be aiming for. One of these is 'caring' which can involve making a difference to the life of others. In Mathematics, this can mean helping others in your Mathematics class with their work, being sympathetic if your classmates are having personal or work difficulties and supporting them all you can.
Carrying capacity (AHL)	(see **logistic function**) (Functions)
Cartesian coordinates	The numbers signifying the position of a point on **2D** (x and y) or **3D** (x, y, and z) **axes**. (Functions)
Cartesian form (of complex number) (AHL)	A complex number z is given in Cartesian form when expressed as $z = x + iy$, corresponding to a point with **Cartesian coordinates** (x, y), where x is the real part and y the imaginary part of z. (Number and algebra)
Cartesian plane	The graph on which **coordinates** are plotted. (Geometry and trigonometry)
Categorical variable	An amount in a category represented by a letter such as x, that can take several values. For example, when describing blood types, the values are A, B, AB and O. (Statistics and probability)
Categorize (categorization) (AHL)	To put data into distinct groups, for example, in preparation for doing a **chi-squared test for independence**. (Statistics and probability)
Causation (cause)	When one outcome is produced as a result of another. For example, eating excessive amounts of chocolate causes obesity. (Statistics and probability)
Cell (on Voronoi diagram)	An area which shows all points that are nearer to a given **site** than any other specified site in a Voronoi diagram. Also called a region or a face. (Geometry and trigonometry)
Census	A survey to gather information from an entire population. (Statistics and probability)
Centi-	A hundredth or 1/100. For example, one centimetre is a hundredth of a metre. (Number and algebra)
Central limit theorem (AHL)	The fact that the distribution of the **mean** of a large number of random samples taken from the same population will approach a **normal distribution**. This is important because it implies for a large population and a sufficiently large sample (for your exam that is $n > 30$) that the mean of the sample will be a good approximation of the population mean. (Statistics and probability)
Central tendency, measure of	Central tendency relates to the likelihood that values of a random variable will cluster around the **mean**, **median**, or **mode**. When we refer to the measure of central tendency, these are most often called **averages**. (Statistics and probability)
Certain event	An event which will definitely happen – it has a probability of 100% of happening. (Statistics and probability)

C

Chain rule (AHL) — The method for finding the **derivative** of **composite functions**.

To find the derivative of $y = f(g(x))$, substitute $u = g(x)$, so $y = f(u)$, then $\frac{dy}{dx} = \frac{dy}{du} \times \frac{du}{dx}$. For example, to find the derivative of $y = \sin(2x)$, let $u = 2x$, then $y = \sin(u)$ and $\frac{dy}{dx} = \frac{d\sin(u)}{du} \times \frac{d(2x)}{dx}$, which gives $\frac{dy}{dx} = \cos(u) \times 2 = 2\cos(2x)$. The chain rule is given in the formula booklet so you do not need to memorize it. (Calculus)

Chance — The possibility that something might occur. (Statistics and probability)

Characteristic polynomial (AHL) — In the process of finding the **eigenvalues** and **eigenvectors** of a linear transformation (that is expressed in terms of a square matrix A), one has to find the values of λ that satisfy the matrix equation $Av = \lambda v \Rightarrow (A - \lambda I)v = 0$. Since the vector v must be non-zero, then the determinant of matrix $(A - \lambda I)$ must be equal to 0. The characteristic polynomial of matrix A is then the polynomial in terms of λ that one arrives at using $|A - \lambda I| = 0$. The solutions of this polynomial are the eigenvalues of matrix A. (Number and algebra)

(see **diagonalization**)

Chinese postman problem (AHL) — A typical problem approached using **graph theory** that supposes there is a postal worker who needs to minimize walking to deliver the post. The postal worker must start and end in the same place and visit each street at least once. It is an example of finding the shortest route through an undirected weighted graph that goes along each edge at least once. (Geometry and trigonometry)

Chinese postman problem algorithm (AHL) — An algorithm to solve the **Chinese postman problem**. As part of your IB studies, you need to be able to apply the algorithm to a weighted graph with up to four odd vertices. You should be able to explain and apply the problem and the algorithm and know why it works. (Geometry and trigonometry)

Chi-squared distribution — A probability distribution used in many **hypothesis tests**. In technical terms, it is the distribution of the sum of the squares of independent random variables. (Statistics and probability)

Chi-squared goodness of fit test — A method to see whether something, eg a **die**, is fair. (Statistics and probability)

Chi-squared statistic — A figure which is calculated when doing a **chi-squared goodness of fit test** or a **chi-squared test for independence**. (Statistics and probability)

Chi-squared test for independence — A statistical test to see whether variables are independent of each other. (Statistics and probability)

Chord — A straight line which goes from one point on a circle to another on the circle. A chord that goes through the centre of the circle is called a **diameter**). (Geometry and trigonometry)

Circuit (AHL) — A path through a graph that begins and ends in the same place. A circuit may revisit vertices. (Geometry and trigonometry)

Circuit graph (AHL) — A graph that forms a **circuit**. (Geometry and trigonometry)

Circular motion (AHL) — Movement along the **circumference** of a circle or following a circular path. (Geometry and trigonometry)

Circumference — The distance around the outside of a circle. (Geometry and trigonometry)

Class — A set or group of data. (Statistics and probability)

Class boundaries — The upper and lower ends of a set of data. (Statistics and probability)

Class interval *or* class width	The difference between the upper and lower boundaries of a set of data. (Statistics and probability)
Classify (classification)	To put into groups according to categories. (Statistics and probability, Calculus)
Closed walk (AHL)	A **walk** through a graph that begins and ends in the same place, ie the first and last **vertex** are the same. (Geometry and trigonometry)
Coefficient	The number in front of a letter in algebra. For example, $5x$ means the variable x is multiplied five times, so the coefficient is 5. (Number and algebra)
Coefficient matrix (AHL)	The **matrix** formed by the **coefficients** of the **variables** in a **system of linear equations**. A system of two linear equations with unknown variables x and y is written as $$\begin{cases} a_{11}x + a_{12}y = b_{11} \\ a_{21}x + a_{22}y = b_{21} \end{cases}$$ where $a_{11}, a_{12}, a_{21}, a_{22}, b_{11}, b_{21}$ are real numbers. The system is written in matrix form as $$\begin{pmatrix} a_{11} & a_{12} \\ a_{21} & a_{22} \end{pmatrix} \begin{pmatrix} x \\ y \end{pmatrix} = \begin{pmatrix} b_{11} \\ b_{21} \end{pmatrix}$$ or $$AX = B$$ where $$A = \begin{pmatrix} a_{11} & a_{12} \\ a_{21} & a_{22} \end{pmatrix}$$ is the coefficient matrix. (Number and algebra) (see **variable matrix, constant matrix**)
Coefficient of determination (AHL)	A measure of how closely the **regression** curve reflects the data, it is written and often called R^2. It represents the proportion of the variance of the dependent variable that is explained by the model. It is a number between 0 and 1, the closer the value is to 1, the closer the model is to the data. In a **linear model**, R^2 is the square of the **Pearson's product-moment correlation coefficient**. It is important to remember that many different factors affect a model so only looking at R^2 is not enough to choose a model. $R^2 = 1 - \dfrac{SS_{res}}{SS_{tot}}$, where SS_{res} is the sum of square residuals and SS_{tot} is the total sum of squares. $SS_{tot} = \sum_{i=1}^{n}(y_i - \bar{y})^2$ $SS_{res} = \sum_{i=1}^{n}(y_i - f(x_i))^2$ (Statistics and probability)
Collection of data	The gathering of information for statistical analysis. You can gather it yourself or get it from past surveys (possibly on the internet). (Statistics and probability)
Collinear (AHL)	Points that lie on the same line. (Geometry and trigonometry)

C

Column (of a matrix) (AHL)

A vertical line of **elements** in a matrix. For example:

$$\begin{pmatrix} 1 & 2 & 3 \\ 3 & 0 & 1 \\ 0 & 1 & 2 \end{pmatrix} \quad \begin{pmatrix} 0 \\ 1 \\ 2 \end{pmatrix} \quad \begin{pmatrix} 0 & 1 & -1 \end{pmatrix}$$

In some cases, a column may consist of only one element. In general, a matrix of the form:

$$A = \begin{pmatrix} a_{11} & a_{12} & \cdots & a_{1n} \\ a_{21} & a_{22} & \cdots & a_{2n} \\ \vdots & \vdots & \ddots & \vdots \\ a_{m1} & a_{m2} & \cdots & a_{mn} \end{pmatrix}$$

column 1   column 2   ... column n

has columns with m **elements**. (Number and algebra)

(see **row (of a matrix)**)

Column representation (AHL)

When a **vector** is represented as a column matrix. For a vector $v = v_1 i + v_2 j + v_3 k$, the column representation is: $\begin{pmatrix} v_1 \\ v_2 \\ v_3 \end{pmatrix}$

(Geometry and trigonometry)

Column state matrix (AHL)

A **matrix** showing the probability of being in each state of a **Markov chain** after a number of transitions. It is written s_n and found by calculating $s_n = T^n s_0$, where n is the number of transitions, T is the transition matrix and s_0 is the initial state matrix. For example, for a Markov chain with transition matrix $\begin{pmatrix} 0.3 & 0.7 \\ 0.6 & 0.4 \end{pmatrix}$ and initial state matrix $\begin{pmatrix} 0.2 \\ 0.8 \end{pmatrix}$, $s_3 = \begin{pmatrix} 0.3 & 0.7 \\ 0.6 & 0.4 \end{pmatrix}^3 \begin{pmatrix} 0.2 \\ 0.8 \end{pmatrix} = \begin{pmatrix} 0.5318 \\ 0.5156 \end{pmatrix}$

(Statistics and probability)

Column vector (AHL)

A **matrix** consisting of m **rows** and one **column**.

$$\begin{pmatrix} a_1 \\ a_2 \\ \vdots \\ a_m \end{pmatrix}$$

For example, a 3×1 column vector, such as $\begin{pmatrix} 2 \\ -1 \\ 4 \end{pmatrix}$ can represent the position of a fly in a 3-dimensional space.

(Number and algebra)

(see **row vector**)

Command term

Words that are frequently used in exam questions and which give you the instruction for what you need to do. For example, **draw**, **calculate**. If you understand the command term in the IB Diploma, you will find it a lot easier to answer the question.

Comment

To give your opinion on something that you have been shown. In Mathematics, this might be a statement or the answer to a calculation. Your opinion might involve giving an explanation which reveals where the mathematical statements are either right or wrong. Can be used as a **command term** at IB Diploma.

Common difference	In **arithmetic sequences**, the amount added on to each term to get to the next term. For example, in the sequence $1, 5, 9, 13, 17$ the common difference is 4. (Number and algebra) (see **constant difference**)
Common ratio	In **geometric sequences**, the amount that each term is multiplied by to get to the next term. For example, in the geometric sequence $1, 4, 16, 64, 256$ the common ratio is 4. (Number and algebra) (see **constant ratio**)
Communicator	The IB learner profile suggests ten qualities that students should be aiming for. One of these is 'communicator' which describes someone who can express views clearly and put them across to others. In Mathematics, you could explain methods to others in your class or explain to your teacher how you managed to work something out.
Commutative (commutativity) (AHL)	The property stating that an **operation** on two or more objects is independent of the order of the objects. Both addition and multiplication of real or complex numbers are commutative: $a + b = b + a$ $a \times b = b \times a$ However, in general, number subtraction and division are not commutative: $a - b \neq b - a$ $a \div b \neq b \div a$ Although the commutative property of addition and multiplication seems obvious for numbers, it may differ for other sets of objects, such as **matrices** or **vectors**. Matrix addition is commutative: $\boldsymbol{A + B = B + A}$ but matrix multiplication in general is not: $\boldsymbol{AB \neq BA}$ Both vector addition and scalar product are commutative: $v + w = w + v$ $v \cdot w = w \cdot v$ but the **vector product** is not: $v \times w = -w \times v$ (Number and algebra, Geometry and trigonometry)
Compare (comparison)	To look at two sets of data, shapes, etc and see how they are similar and how they differ. Can be used as a **command term**.
Compare and contrast	Used as a **command term** during your IB Diploma studies. You look at two (or more) items or situations and describe or give details about them. Your description must cover the similarities and the differences between these items. Take care to refer to both (or all) of the items/situations throughout your work.
Compass direction	The main compass directions are north, south, east, and west. (Geometry and trigonometry) (see **bearing**)
Complement	Used in sets. We use the symbol ' to denote the complement of a set, so A' is the complement of set A and consists of the elements which are NOT in set A. (Statistics and probability)

C

Complementary events — A pair of results of the form result A and result 'not A'. For example, when tossing a coin, heads and tails are complementary events because if you don't throw heads ('not heads') you throw tails. (Statistics and probability)

Complete graph (AHL) — A graph where each pair of **vertices** is connected by an **edge**. (Geometry and trigonometry)

Complex (AHL) — (see **complex number** or **complex quantity**) (Number and algebra)

Complex conjugate (AHL) — (see **conjugate (of a complex number)**) (Number and algebra)

Complex number (AHL) — A number that can be written as the sum of a real number and an imaginary number, ie a multiple of $i = \sqrt{-1} \Leftrightarrow i^2 = -1$. (Number and algebra)

Complex plane (AHL) — Using the **Cartesian plane** to represent **complex numbers** of the form $z = x + iy$ as points with coordinates (x, y) or position vectors $\binom{x}{y}$. (Number and algebra) (see **Argand diagram**)

Complex quantity (AHL) — A quantity that can be expressed by a **complex number**. (Number and algebra)

Component (of a vector) (AHL) — The parts of a vector. For example, given the vector $v = 3i + 2j + 4k$ the x-component of the vector is 3, the y-component is 2 and the z-component is 4. (Geometry and trigonometry)

Composite function (AHL) — Given two functions $f: A \to B$ and $g: B \to C$, their functional composite $g \circ f: A \to C$ is defined such that it relates every $x \in A$ to the image of $f(x)$ through g. Note that $(g \circ f)(x)$ is sometimes written as $g(f(x))$.

For example, if $f(x) = x^2$ and $g(x) = 5x - 2$, then $(g \circ f)(x) = 5x^2 - 2$ whereas $(f \circ g)(x) = (5x - 2)^2$.

So, $(g \circ f)(-2) = 5 \cdot 4 - 2 = 8$

and $(f \circ g)(3) = (13)^2 = 169$.

For $g \circ f$ to be defined properly, the domain of g must be identical to the range of f. If g includes **domain restrictions**, then it is likely that f also needs to be restricted, so that its range is equal to the domain of g. Note this could be a possible type of exam question. For example, if $f(x) = 5x - 2$ and $g(x) = \sqrt{x}$, the (restricted) domain of g is $x \geq 0$. Therefore, the domain of f must be restricted so that its range is $y \geq 0 \Rightarrow 5x - 2 \geq 0 \Rightarrow x \geq \frac{2}{5}$. (Functions)

Composite transformation (AHL) — The function obtained as a result of more than one **transformation** taking place. For example, when the function $y = x^3$ is transformed by a vertical stretch of scale factor 2 followed by a vertical translation of 5, the resulting function from this composite transformation is $y = 2x^3 + 5$. (Functions)

Composition of transformations (AHL) — A combination of two or more **transformations**. (Geometry and trigonometry)

Compound interest — A method of calculating interest where the interest is calculated on the amount borrowed plus the sum of any previous interest. It can be calculated annually, half yearly, quarterly, monthly, etc. (Number and algebra)

Compound shape — The result of two or more shapes being fitted together. (Geometry and trigonometry)

Compounded growth — How an amount has grown, when the increase uses **compound interest**. (Number and algebra)

Compute (computation)	To work out – not necessarily with a computer.
Computer notation	A way of writing something that is used on a computer or calculator, but should not be used in written work. For example, **standard form** is written 5.6E7 on a calculator or computer but should be written as 5.6×10^7 in your own work.
Concave down (AHL)	A curve that bends downward. A graph (or a function) is concave down at a point when the curve is below the **tangent** line at that point. The **second derivative** of the function is less than 0 at that point, $f''(x) < 0$. (Calculus)

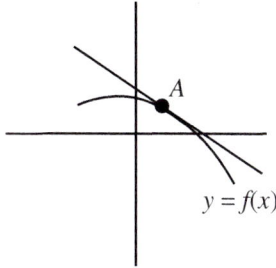

Concave up (AHL)	A curve that bends upward. A graph (or a function) is concave up at a point when the curve is above the **tangent** line at that point. The **second derivative** of the function is greater than 0 at that point, $f''(x) > 0$. (Calculus)

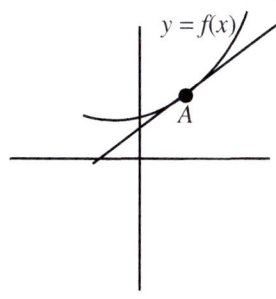

Concept	An idea. Remember that if you understand a concept, you do not have to memorize formulae, etc. So if you understand the concept of **Pythagoras' theorem**, you can calculate any side in any right angled triangle.
Conclude (conclusion)	A comment on the result of an enquiry or question. In an **IA** it will be what the overall results show and whether a **hypothesis** is true or false.
Condition	A requirement that is needed in order for something to be true.
Conditional probability	The probability of one event happening given that another event has already occurred. (Statistics and probability)
Cone	A **three-dimensional** shape with a circle as a base and a point at the top. (Geometry and trigonometry)
Confidence interval (AHL)	The **range** of values between which the true value of a population statistic, eg **mean**, lies given a specified **probability**. For example, given a sample 1, 2, 3, 4, 5 from a normally distributed population; the mean of the population has a 95% confidence interval of [1.76, 4.24]. For a **t-distribution**, a confidence interval is found using the formula $\bar{x} \pm t \times \frac{s_{n-1}}{\sqrt{n}}$ For a **normal distribution**, a confidence interval is found using the formula $\bar{x} \pm z \times \frac{\sigma}{\sqrt{n}}$ Both formulae are given in the formula book. (Statistics and probability)

C

Confidence level (AHL)	The specified **probability** that the true value is contained within the **confidence interval**. (Statistics and probability)
Congruent	Exactly the same in every way (sides and angles) – often used for triangles. (Geometry and trigonometry)
Conic section	A conic section is where a surface of a **cone** meets with a **plane** and forms a curve. (Functions)
Conjugate (of complex number) (AHL)	The **complex number** $z^* = x - iy$ is the conjugate of complex number $z = x + iy$. It can be easily derived that $z \cdot z^* = |z|^2$, where $|z|$ is the **modulus** of both complex number z and its conjugate z^*. It is also true that $z + z^* = 2x = 2\text{Re}(z) = 2\text{Re}(z^*)$ and $z - z^* = 2iy = 2i\text{Im}(z) = -2i\text{Im}(z^*)$, where $\text{Re}(z)$ is the **real part** of z and $\text{Im}(z)$ denotes its **imaginary part**. (Number and algebra)
Conjugate pair (AHL)	A **complex number** z together with its **conjugate** z^*. When a **polynomial equation** with real-number coefficients accepts a complex number z as its solution, then the conjugate of z, denoted by z^*, is also a solution of the equation. Then the polynomial expression accepts $x - z$ and $x - z^*$ as linear factors, or $x^2 - 2\text{Re}(z) \cdot x + |z|^2$ as a quadratic factor. Note that the term conjugate pair is also used to describe pairs of real numbers of the form $2 + \sqrt{3}$ and $2 - \sqrt{3}$. (Number and algebra)
Connected graph (AHL)	A graph in which there is a **path** connecting any **vertex** to any other vertex. (Geometry and trigonometry)
Consecutive	One after the other. For example, consecutive even numbers would be 2, 4, 6, 8. (Number and algebra)
Consequently	As a result of something happening, another event happens.
Constant	Something which does not change, usually just a number. For example, in $y = 5x + 3$, 3 is the constant. We also might refer to a 'constant speed' where the pace at which something travels remains the same. (Number and algebra, Functions) (see **variable**)
Constant difference	Another term for **common difference**. (Number and algebra)

18

Constant matrix (AHL)	A matrix in which all **entries** are **constants,** usually numbers. A constant matrix is often used when writing **systems of linear equations** in matrix form. A system of two linear equations with unknown variables x and y is written as: $$\begin{cases} a_{11}x + a_{12}y = b_{11} \\ a_{21}x + a_{22}y = b_{21} \end{cases}$$ where $a_{11}, a_{12}, a_{21}, a_{22}, b_{11}, b_{21}$ are real numbers. The system is written in matrix form as: $$\begin{pmatrix} a_{11} & a_{12} \\ a_{21} & a_{22} \end{pmatrix} \begin{pmatrix} x \\ y \end{pmatrix} = \begin{pmatrix} b_{11} \\ b_{21} \end{pmatrix}$$ or $$\boldsymbol{AX = B}$$ where $$\boldsymbol{B} = \begin{pmatrix} b_{11} \\ b_{21} \end{pmatrix}$$ is the constant matrix. (Number and algebra) (see **coefficient matrix, variable matrix**)
Constant ratio	A **ratio** that does not change. (Number and algebra) (see **common ratio**)
Constant term	Something that does not change. For example, in $y = 2x + 3$, 3 is the constant term. (Calculus)
Constraint	A condition that restricts what we can do. When we are using **optimization** in **calculus**, a restriction can be put on what we need to do. For example, the length must not be more than twice as long as the width. (Calculus)
Construct (construction)	To draw accurately, usually using a ruler and compasses, eg when you draw a triangle, given the lengths of all three sides. Often used as a **command term**. (Geometry and trigonometry)
Content validity (AHL)	A measure of how accurately a **survey** is able to obtain the information it is trying to assess. For example, for a questionnaire, content validity is a measure of how representative the chosen questions are of all the possible questions that could be asked. (Statistics and probability)
Context	The general situation to which something relates and which helps us understand it. When solving real-life problems, we need to know the setting (or context) in which the problem occurs. For example, if we are solving a quadratic equation and one of the answers is negative, we can discount it if we are trying to find a **length**, as we know lengths cannot be **negative**.
Contingency table	Data shown in a tabular form. The tables are used in **chi-squared tests for independence** to show observed data before testing whether two variables are independent of each other. (Statistics and probability)
Continuity (of a function) (AHL)	The property of whether a function is **continuous** at a point or over a specified range. (Functions) (see **continuous function**)
Continuous (data)	Something which can be measured, and is not restricted to whole numbers, so has an infinite number of possible values. (Statistics and probability) (see **discrete data**)

C

Continuous (function) (AHL) — A function f is continuous at a point x_0, if the **limit** of f exists at x_0 and satisfies $\lim_{x \to x_0} f(x) = f(x_0)$. This means that f must be defined at x_0, and at points infinitesimally close to x_0 the function takes values converging to $f(x_0)$. The points at which a function is not continuous are called discontinuities. A function that is continuous in all points of its domain is called continuous (in its domain). (Functions)

Continuous compounding — The maximum sum that can theoretically be obtained when **compound interest** is reinvested on a continual basis (rather than on a periodic basis such as every month). (Number and algebra)

Convenience sampling — A way of gathering data by just taking the easiest path or the population which is easily available, for example, by using the people in your class. (Statistics and probability)

(see **bias**)

Converge (AHL) — The sum of all terms of a decreasing **geometric sequence** (with **common ratio** $|r| < 1$) converges or approaches to $S_\infty = \frac{u_1}{1-r}$. It may be counter-intuitive that a sum with infinite terms has a finite value, however, it should be considered as the **limit** of the sum of the first n terms of the sequence as n approaches infinity, $S_\infty = \lim_{n \to \infty} \sum_{k=1}^{n} u_k$.

This sum converges to a finite value because of the decreasing nature of the terms of such a sequence, ie $\lim_{n \to \infty} u_n = 0$. (Number and algebra)

(see **infinite geometric series**)

Convergent (convergence) — Moving towards a limit. (Number and algebra, Calculus)

Converse — To change the order of a statement, not necessarily keeping its meaning the same. For example:

'all multiples of 6 are multiples of 2' (true)

But the converse of this is:

'all multiples of 2 are multiples of 6' (not true)

Conversion graph — A graph which is used to convert from one unit to another, for example, from kilometres to miles. (Functions)

Convert (conversion) — To change from one unit to another. For example, to convert centimetres into metres you divide by 100.

Convex hull — Used in work on **Voronoi diagrams**, the convex hull of a set of x points, is the smallest shape (with all angles less than 180 degrees) in which those x points will fit. (Geometry and trigonometry)

Convex polygon — A many sided **2D** shape, where all interior angles are less than 180 degrees, eg a regular pentagon where all angles are 108 degrees. (Geometry and trigonometry)

Coordinates — The numbers signifying the position of a point, written as (x, y). (Geometry and trigonometry)

Correct to (2 decimal places, 3 significant figures, etc) — Estimating to a certain degree of accuracy.

IB Diploma questions will usually specify 'correct to 3 significant figures'. For example:

8540 is 8536 correct to 3 significant figures

0.00654 is 0.006542 correct to 3 decimal places. (Number and algebra)

Correlation (correlated) — A relationship between two variables so that when one changes, so does the other. (Statistics and probability)

Correlation coefficient	This is a value between −1 and 1 to show the strength of the relationship between two sets of data. (Statistics and probability)
	(see **Pearson's product-moment correlation coefficient**)
Cosine (cos) function	A trigonometric function. The cosine of an acute angle can be between 0 and 1 and a graph can be drawn to represent it. (Functions)
	(see **cosine ratio**, **sine (sin) function**, **tangent (tan) function**)
Cosine (cos) ratio	In a **right-angled triangle**, the length of the **adjacent** side divided by the length of the **hypotenuse**. (Geometry and trigonometry)
Cosine rule	A method used to find an angle or side in a non-right-angled triangle. The formulae are given on your formula sheet, according to whether you are looking for a side or an angle. (Geometry and trigonometry)
Cost function, $C(x)$	Provides the cost of a product or service as a function of the quantity of an ingredient used or the number of items produced, x, eg it can be given by functions like $C(x) = 5x - 4$, or $C(x) = -0.02x^2 + 5x + 20$, etc. (Functions)
Counter-example	Something that proves that a statement is not true. For example, if a statement is made that all prime numbers are odd, 2 would be the counter-example that proves this is untrue.
Coupled differential equations (*or* coupled system) (AHL)	A set of two **differential equations** containing three variables, two of which depend on the third. For example: $$\frac{dy}{dt} = 2y + x$$ $$\frac{dx}{dt} = 5x + 4y$$ where x and y depend on t.
	Coupled differential equations are used to model movements of a population, such as predator-prey situations. (Calculus)
Criterion-related validity (*or* criterion validity) (AHL)	A measure of how well the results of a **survey** are able to predict what they are trying to predict. For example, if a group of IB Diploma Mathematics students sit a mock exam and the results of the mock exam can accurately predict their results in the final exam, the mock exam is said to have criterion-related validity. (Statistics and probability)
Critical region	The values in **hypothesis testing** where you can reject the **null hypothesis**. (Statistics and probability)
Critical value	In a hypothesis test, a number which is compared to a test statistic to determine if the **null hypothesis** should be rejected. For example, in a **chi-squared test for independence** if the calculated chi-squared statistic is greater than the critical value, the null hypothesis is rejected. (Statistics and probability)
Cross product (AHL)	(see **vector product**) (Geometry and trigonometry)
Cross-section	When you cut through a **3D** shape in a certain way, the cross section is the face which appears all the way across the shape. (Geometry and trigonometry)
Cube	1 The cube of a number is the number produced when you multiply it by itself twice. For example:
	The cube of 4 is 64.
	(Number and algebra)
	2 A **three-dimensional** shape with six faces, which are all square.
	(Geometry and trigonometry)

C

Cube number	A number to the power of 3 or multiplied by itself twice. For example: 4 cubed is $4^3 = 4 \times 4 \times 4 = 64$. (Number and algebra)
Cube root	A number which when multiplied by itself twice, gives you a number of which it is the cube root. For example: the cube root of 64 is 4 as $4 \times 4 \times 4 = 64$. (Number and algebra)
Cubic (metre, centimetre, millimetre)	A unit we use to measure volume. (Number and algebra)
Cubic function	Where the highest power of x is x^3. For example: $y = 5x^3 + 3x^2 + 7x - 4$ (Functions)
Cubic graph (AHL)	A graph where the degree of every **vertex** is 3. (Geometry and trigonometry)
Cubic model	A **function** used to describe a real-life situation where the function is cubic. For example, the volume of a **3D** object. (Functions) (see **cubic function**)
Cubic regression (AHL)	A **regression** curve (curve of best fit) that takes the form of a **cubic function**. (Statistics and probability)
Cuboid	A **three-dimensional** shape with six faces which are all rectangular. (Geometry and trigonometry)
Cumulative	The total so far; adding up all the values as you go along, making a new total each time. (Statistics and probability)
Cumulative frequency	Adding up all the values up to and including the value given. For example, if 0 – 10 has 12 values, 10 – 20 has 10 values, and 20 – 30 has 8 values, the cumulative frequency would be up to 10, 12 up to 20, 12 + 10 = 22, and up to 30, 12 + 10 + 8 = 30 (Statistics and probability)
Cumulative frequency curve	A line on a graph which represents **cumulative frequency** where the points are joined in an S shaped line (or 'curve'). (Statistics and probability)
Cumulative frequency graph	A graph which represents **cumulative frequency** where the points are joined in an S shaped line (or 'curve'). (Statistics and probability)
Currency	The currency of a country is the money which it uses. For your IB Diploma studies, currency is usually given as three letters, for example, USD for United States dollars. (Number and algebra)
Cycle (AHL)	A **path** in a graph that begins and ends in the same place without revisiting any **vertices** (other than the start and end). (Geometry and trigonometry)
Cycle algorithm (AHL)	An **algorithm**, or set of steps, used to find cycles in a graph. (Geometry and trigonometry)
Cylinder	A **three-dimensional** shape with a circle as its base, as its top and as its **cross-section**. (Geometry and trigonometry)

Data (*singular* datum)	Information which has been collected from somewhere and may involve facts, numbers, or measurements. (Statistics and probability)
	(see **qualitative**, **quantitative**)
Data analysis	The process of categorizing, organizing, and commenting on the information collected. (Statistics and probability)
Data collection	The process of obtaining information; this may be done by questioning people, observation, or measurement, or we may use information from a **survey** already available on the internet, for example. (Statistics and probability)
	(see **census**, see **sample**)
Data set (*or* dataset)	The information collected from a **survey** or from other observations or measurements; for example, the heights of 100 people. (Statistics and probability)
Data source	The place from where information is obtained. Some sources are more reliable than others. For example, the World Health Organization website is generally thought to be reliable, but a source such as Wikipedia less so. (Statistics and probability)
De Moivre's theorem (AHL)	Provides an easy way to calculate the value of integer powers of **complex numbers**. More specifically, using the **polar form** of complex numbers, the theorem states that $(\cos x + i \sin x)^n = \cos(nx) + i \sin(nx)$ where $n \in \mathbb{Z}$ and $x \in \mathbb{R}$. (Number and algebra)
Debt	An amount, usually money, which somebody owes to someone else. (Number and algebra)
Decay	When a **function** is decreasing. (Functions)
	(see **exponential decay**)
Decimal	Based on units of 10. A decimal number is one which includes a decimal point. For example, 1.5 represents 1 and 5 tenths. (Number and algebra)
Decimal expansion	A way of showing how numbers are made up in base 10. For example: 243 is $2 \times 10^2 + 4 \times 10^1 + 3 \times 10^0$ (Number and algebra)
Decimal number system	Based on 10 and its powers – called hundreds, tens and units at its basic level. (Number and algebra)
Decimal place	As counted from the **decimal point**. For example, 5.43 is given to two decimal places as there are two numbers after the decimal point. (Number and algebra)
Decimal point	A point (or small dot) which separates whole numbers from amounts less than a whole. For example, 5.7 is the same as 5 and 7 tenths. (Number and algebra)
Decode (a message) (AHL)	Understand how to read a text message which is written in an encrypted form. (Number and algebra)
	(see **encode (a message)**)
Decreasing function	A graph which has a negative gradient. (Calculus)
Deduce (deduction)	To work out from information given. Used as a **command term** in your IB studies.
Default	The preset selection of technology. When you reset your **GDC**, it returns to its default setting, or the settings it had when it was delivered to you. At SL, you must remember that this includes **radians**, which are not used for some exams. You can select **degrees** from the mode button.

D

Definite integral — When you are integrating and are given limits between which you must integrate. (Calculus)

(see **integral**)

Degree — A measure of angles. There are 360 degrees in a full rotation. The symbol for degrees is °. (Geometry and trigonometry)

Degree (of a graph) (AHL) — The sum of the degrees of the **vertices** of a graph. (Geometry and trigonometry)

Degree (of a polynomial) — The highest power of x in an expression. For example, $y = x^3 + 7x^2 - 3$ has a degree of 3. (Number and algebra)

Degree (of a vertex) (AHL) — The number of **edges** that join that **vertex**. (Geometry and trigonometry)

Degree mode (*or* degrees mode) — The mode in which your calculator should be in, unless you are using **radians** for angles. Check on your mode before you start trigonometry as you will get the wrong answer if it is set to radians.

(see **radian mode**)

Degree of accuracy — A measure of how close or precise a value is to the actual or real value, so shows the **error** involved in an **approximate** calculation. (Number and algebra)

Degrees of freedom — A value in a chi-squared calculation of (number of rows − 1) x (number of columns − 1). This is used to decide on the critical value which can be read from a table but is usually given to you in a question. (Statistics and probability)

Delete (deletion) — To remove or take away. (Geometry and trigonometry)

Deleted vertex algorithm (AHL) — An algorithm used to find the **lower bound** in the **travelling salesman problem**, ie it finds the smallest distance that the solution could be. (Geometry and trigonometry)

Demonstrate — To show or explain why something is as it is. Often used as a **command term** in your IB studies, where you must back up your work with evidence and reason which also should include examples and/or practical application.

Denominator — The bottom part of a fraction which shows the number of parts the whole is divided into. For example, in ½, the 2 is the denominator. (Number and algebra)

Denote — To show or indicate.

Dependence — Reliance or relationship. For example, we might talk about a person's weight being dependent on their height. (Statistics and probability)

Dependent events — If the occurrence of one event affects the occurrence or outcome of another, the two events are said to be dependent. (Statistics and probability)

(see **independent events**)

Dependent variable — Something which is reliant on something else. (Statistics and probability)

Depreciation — When the value of something goes down over time, usually expressed as a percentage. (Number and algebra)

(see **annual depreciation**)

Derivative — The result obtained after **differentiation**. (Calculus)

Derive (derivation)	If you say something derives from something else, you are describing its source or where it comes from. In Mathematics, we may use this as an informal term for **differentiation**.
Derived unit	Words for measurements which are based on the basic units. For example, newtons are derived from kg per square metre. (Number and algebra)
Descending order	In order from the highest number to the lowest. (Number and algebra) (see **ascending order**)
Describe (description)	Providing a detailed account so that something can be visualized. For example, we may describe the properties of a pentagon. Used as a **command term** in your IB studies.
Descriptive statistics	Describing or summarising information. This can be a collection of data, from which we can find mean, median, mode, etc. (Statistics and probability)
Determinant (of a square matrix) (AHL)	In a 2×2 matrix $$A = \begin{pmatrix} a & b \\ c & d \end{pmatrix}$$ the determinant, denoted by $|A|$ or $\det A$, is a real number defined as: $$|A| = \det A = \begin{vmatrix} a & b \\ c & d \end{vmatrix} = ad - bc$$ In a 3×3 matrix $$A = \begin{pmatrix} a & b & c \\ d & e & f \\ g & h & i \end{pmatrix}$$ the determinant is a real number defined as: $$|A| = \det A = \begin{vmatrix} a & b & c \\ d & e & f \\ g & h & i \end{vmatrix} = a\begin{vmatrix} e & f \\ h & i \end{vmatrix} - b\begin{vmatrix} d & f \\ g & i \end{vmatrix} + c\begin{vmatrix} d & e \\ g & h \end{vmatrix}$$ The first row of the matrix is expanded in the example above. The same calculation can be done by expanding along any row or column. For example, expanding the following determinant along the first column yields: $$\begin{vmatrix} -1 & 2 & 5 \\ 0 & 4 & -1 \\ 9 & 3 & 4 \end{vmatrix} = -1 \times \begin{vmatrix} 4 & -1 \\ 3 & 4 \end{vmatrix} - 0 \times \begin{vmatrix} 2 & 5 \\ 3 & 4 \end{vmatrix} + 9 \times \begin{vmatrix} 2 & 5 \\ 4 & -1 \end{vmatrix}$$ $$= -[4 \times 4 - (-1) \times 3] + 9[2 \times (-1) - 5 \times 4]$$ $$= -217$$ By expanding along the second row the result is the same: $$\begin{vmatrix} -1 & 2 & 5 \\ 0 & 4 & -1 \\ 9 & 3 & 4 \end{vmatrix} = -0 \times \begin{vmatrix} 2 & 5 \\ 3 & 4 \end{vmatrix} + 4 \times \begin{vmatrix} -1 & 5 \\ 9 & 4 \end{vmatrix} - (-1) \times \begin{vmatrix} -1 & 2 \\ 9 & 3 \end{vmatrix}$$ $$= 4 \times [(-1) \times 4 - 5 \times 9] + 1 \times [(-1) \times 3 - 2 \times 9]$$ $$= -217$$ (Number and algebra)

D

Determine — To work out the only possible answer from information given. A **command term** used in your IB studies.

Diagonal (of a matrix) (AHL) — All **elements** $a_{11}, a_{22}, a_{33}, \ldots, a_{nn}$ that lie on the diagonal line running from the top left to the bottom right of a **square matrix**. For example:

$$\begin{pmatrix} 2 & 1 \\ 5 & 0 \end{pmatrix} \qquad \begin{pmatrix} 0 & 3 & 4 \\ 1 & 7 & 0 \\ 6 & -4 & 9 \end{pmatrix}$$

(Number and algebra)

(see **diagonal elements**)

Diagonal elements (AHL) — **Elements** that lie on the **main diagonal** of a matrix, running from top left to bottom right. The position of the diagonal elements a_{ij} is such that the number of the row and column are equal ($i = j$). For example:

$$\begin{pmatrix} a_{11} & a_{12} & a_{13} & \cdots \\ a_{21} & a_{22} & a_{23} & \cdots \\ a_{31} & a_{32} & a_{33} & \cdots \\ \vdots & \vdots & \vdots & \ddots \end{pmatrix}$$

(Number and algebra, Geometry and trigonometry)

(see **diagonal (of a matrix)**)

Diagonal matrix (AHL) — A **square matrix** in which all elements outside the **main diagonal** are zero.

$$\begin{pmatrix} 2 & 0 \\ 0 & -1 \end{pmatrix} \qquad \begin{pmatrix} -3 & 0 & 0 \\ 0 & 1 & 0 \\ 0 & 0 & 2 \end{pmatrix}$$

In this case, the diagonal elements represent the **eigenvalues** of the matrix. (Number and algebra)

Diagonalization (AHL) — The conversion of a **square matrix** into an equivalent **diagonal matrix** through a series of basic operations. The **eigenvalues** of the original matrix are then clearly represented by the **diagonal elements** of the diagonal matrix. For example, the diagonalization of the matrix:

$$A = \begin{pmatrix} 4 & 1 \\ -8 & -5 \end{pmatrix}$$ results in the diagonal matrix:

$$D = \begin{pmatrix} 3 & 0 \\ 0 & -4 \end{pmatrix},$$

where 3 and -4 are the **eigenvalues** of matrix A. Then the relation between matrices A and D is given by $A = P \cdot D \cdot P^{-1}$, where matrix P consists of the respective **eigenvectors** of matrix A (each eigenvector being a **column of matrix P**), and P^{-1} is the **inverse of matrix P** (for detailed calculations see the **characteristic polynomial**).

In the above example, $P = \begin{pmatrix} 1 & 1 \\ -1 & -8 \end{pmatrix}$ and $P^{-1} = \frac{1}{7}\begin{pmatrix} 8 & 1 \\ -1 & -1 \end{pmatrix}$. Note that in IBDP exams at HL only cases of 2×2 matrices with two **distinct** eigenvectors will be set. Therefore, matrix P will always have an inverse. (Number and algebra)

Diagrammatic representation — A graph or some other sort of drawing that explains something.

Diameter — The line which goes from one side of a circle to the other, passing through the centre. (Geometry and trigonometry)

Dice (*singular* die) — A cube, usually made of plastic, which has numbers or markings on it and which can be thrown. The most common die is a cube with dots on each face to represent numbers 1–6. Often used in probability questions. (Statistics and probability)

Differential calculus — See **differentiate**. (Calculus)

Differential equation (AHL)	An equation that involves the **derivative** of a function. For example, $\frac{dy}{dx} = 4x^3$ is a differential equation. You will see differential equations in context with rates of change, such as growth of a population. You should know how to form a differential equation from a question given in context. (Calculus)
Differentiate (differentiation)	To use a method where you can find the **derivative** in order to discover, for example, the gradient of a curve at a point. This might sometimes be used as a **command term** in exam questions. (Calculus)
Differentiation from first principles (AHL)	The method of finding the **derivative** of a function using the limit of the **gradient function** as the step to the next point (h) gets smaller and smaller. $$f'(x) = \lim_{h \to 0} \frac{f(x+h) - f(x)}{h}$$ You do not need to know this for the exam. The formula is given in the formula booklet so you do not need to memorize it. (Calculus)
Digit	An individual number. (Number and algebra)
Dimension (in geometry)	The size of something which can be measured in one direction, for example, length, height, width. One dimension (**1D**) is when we measure in one direction, usually length (a line). Two dimensions (**2D**) is when we measure in two directions, usually length and width (eg a square). Three dimensions (**3D**) is when we measure in three directions, usually length, width, and breadth (eg a cube). (Geometry and trigonometry)
Dimension (of a matrix) (AHL)	The number of rows by the number of columns of a **matrix**. A matrix with m **rows** and n **columns** has dimension $m \times n$ (m by n). We may also say that the matrix has an **order** of m by n. For example: $$\begin{pmatrix} 5 \\ 3 \\ 1 \end{pmatrix} \quad (-2 \quad 7) \quad \begin{pmatrix} 0 & 2 \\ -3 & 0 \\ 1 & -4 \end{pmatrix}$$ $3 \times 1 \qquad 1 \times 2 \qquad 3 \times 2$ (Number and algebra)
Direct proportion (or direct variation)	When two variables are linked in a way such that as one increases, so does the other. (Functions) (see **inverse proportion**)
Directed graph (AHL)	A graph where the **edges** are given a direction. (Geometry and trigonometry)
Directed line segment (AHL)	A line segment with a specific given direction, in other words, a **vector**. (Geometry and trigonometry)
Direction field (AHL)	See **slope field**. (Calculus)
Direction vector (AHL)	A vector with a magnitude of 1. It gives the direction in which a point is moving. Also called a unit vector. (Geometry and trigonometry)
Discontinuous (AHL)	A function that is not continuous, ie there is at least one place in the curve where the function does not continue. If you must pick up your pencil to draw different parts of the curve, it is discontinuous. For example, the curve $$y = \begin{cases} x - 2 & x \leq 3 \\ 2x & x > 3 \end{cases}$$ is discontinuous at $x = 3$. (Calculus)

D

Discrete (data) — Data which can only be represented in whole numbers or by a finite number of groups. Examples include the number of brothers and sisters people have, and people's blood group. (Statistics and probability)

Discrete dynamical system (AHL) — A system where state changes occur over time in discrete time steps. (Statistics and probability)

Discrete random variable — Something which can only take a **discrete** value and is the result of a **random sample**. (Statistics and probability)

Dispersion, measure of — In statistics, how far individual values are from the **mean** of the group. (Statistics and probability)

Displacement (AHL) — The difference between an object's current position and its original position. For example, an object that has travelled 10 metres in the positive x-direction and then 15 metres in the opposite x-direction has a displacement of -5 metres in the x-direction. (Calculus)

Displacement function (AHL) — A function that describes the **displacement** of an object at time t. (Calculus)

Displacement vector (AHL) — A **vector** whose magnitude is the straight-line distance between the starting and finishing position. (Geometry and trigonometry)

Distinct (AHL) — Two values that are different from each other. For example:

- a quadratic equation $ax^2 + bx + c = 0$, $a \neq 0$ has two distinct (or different) solutions $x_1 = \dfrac{-b-\sqrt{\Delta}}{2a}$ and $x_2 = \dfrac{-b+\sqrt{\Delta}}{2a}$ when $\Delta > 0$.
  But when $\Delta = 0$, there are two equal (repeated) solutions $x_1 = x_2 = \dfrac{-b}{2a}$.

- a matrix A has distinct (different) eigenvalues (and corresponding eigenvectors) if its characteristic polynomial has distinct roots. In general eigenvalues may be repeated or complex, but in IBDP exams only the calculation of distinct real eigenvalues will be required.

(Number and algebra, Calculus)

Distinguish — To identify and show how two (or more) concepts or items are distinct and different from each other. Often used as a **command term** in your IB Diploma exams.

Distribution — Describes the spread of data. It can be shown graphically, eg on a **normal distribution** histogram. (Statistics and probability)

Distributive (distributivity) (AHL) — The property stating how to **expand** the **brackets** when a quantity is multiplied by the sum (or difference) of two other quantities:

$a(b + c) = ab + ac$

$a(b - c) = ab - ac$

The property can be extended to double brackets:

$(a + b)(c + d) = ac + ad + bc + bd$

The property is valid in both number (real and complex) and matrix sets. (Number and algebra, Geometry and trigonometry)

Division (divide) — One of the four basic operations of Mathematics, which involves sharing an amount, and is represented by the symbol $\div$. (Number and algebra)

Divisor — The number by which you **divide**. (Number and algebra)

Domain	The range of values that can be taken by the **independent variable** (for x and y, this would be x). (Functions) (see **range (of a function)**)
Domain restriction (AHL)	Any real value at which a relationship or function cannot be defined. When defining a function on the maximum possible domain, one has to examine whether there are any real values that need to be excluded from the domain. For example, for $f(x) = \frac{g(x)}{h(x)}$, all x values making $h(x) = 0$ should be excluded from the domain of f; for $f(x) = \ln g(x)$, all values making $g(x) \leq 0$ should be excluded from the domain of f; for $f(x) = \sqrt[n]{g(x)}$ (where n is a positive even integer), all values making $g(x) < 0$ should be excluded from the domain of f; etc. Then the domain is restricted to the remaining real values. The domain of a relationship or function f may be subject to restrictions when one considers the **composite function** $g \circ f$, so that the corresponding range of f is equal to the domain of g. In addition, a function f may need to have its domain restricted in order to become **one to one**, so that its **inverse** relationship becomes a function. For example, the function $f(x) = x^2 - 4x + 7 = (x-2)^2 + 3$, with its vertex at $(2,3)$, is not one to one for all $x \in \mathbb{R}$, thus its inverse relationship is not a function. For its inverse relationship, f^{-1}, to be a function, the domain of f can include only one of the two symmetric parts of the parabola, ie it must be restricted to either $x \leq 2$ or $x \geq 2$ (or any interval that is a subset of these). (Functions)
Dot product (AHL)	(see **scalar product**) (Geometry and trigonometry)
Draw	To make a diagram of something, eg a triangle to work out a trigonometry sum, or a graph of some sort. Can be used as a **command term** in your IB studies. (see **sketch**)
Dual edge	A connection between two vertices on a **dual graph**. (Geometry and trigonometry)
Dual graph	Given a planar graph, G, the dual graph, G*, is a graph that has a vertex for each region of the planar graph and has an edge where two regions of G are separated by an edge. To construct a dual graph, place a vertex in each region of G, including the outside, and connect these vertices when the borders of the corresponding regions have at least one edge in common. For example, given the planar graph shown in black here, its dual graph is shown in green curves. (Geometry and trigonometry) 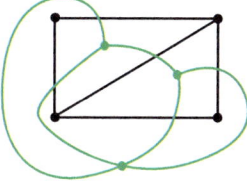
Dynamic geometry software	A program or piece of software that helps you to draw graphs and diagrams, for example, you may use GeoGebra in school. (Geometry and trigonometry)
Dynamic system (or dynamical system) (AHL)	A **mathematical model** that shows how something changes over time. For example, the **exponential growth** of a population or the position of a swinging pendulum can be described as dynamic systems. (Calculus)

E

e	An irrational number used in exponential functions. The first four digits of e are 2.718.... (Number and algebra)
Edge	1 A line on a shape which joins two **vertices**. (Geometry and trigonometry)
	2 (AHL) In graph theory, a connection between **vertices** (points) on a graph. For an example, see **graph**. (Geometry and trigonometry)
	3 (AHL) The lines on a **Voronoi diagram**. They are the set of points that are equidistant (the same distance) from the two closest **sites**. In other words, they are the perpendicular bisectors between two sites. (Geometry and trigonometry)
Edge disjoint (AHL)	When two paths in a graph do not share an edge. (Geometry and trigonometry)
Edge set (AHL)	The set of edges in a graph. (Geometry and trigonometry)
Efficiency (of an unbiased estimator) (AHL)	The extent to which an **unbiased estimator** is an approximation of the population statistic. The efficiency is given by the **variance** of the unbiased estimate. (Statistics and probability)
Eigenvalue (AHL)	When a linear transformation, defined in terms of a **square matrix** A, is applied on a vector v (both of dimension n), producing a multiple of vector v (or else the linear transformation produces a stretch of vector v), then vector v is called an **eigenvector** of the linear transformation that is defined by matrix A, and the scale factor, λ, by which vector v is stretched, is called the eigenvalue of the linear transformation that is defined by matrix A.
	In symbolic form, the above can be represented by the matrix equation:
	$Av = \lambda v$
	(Number and algebra, Calculus)
	(see **characteristic polynomial**)
Eigenvector (AHL)	When a linear transformation, defined in terms of a **square matrix** A, is applied on a vector v (both of dimension n), so that this matrix equation is true,
	$Av = \lambda v$
	then vector v is called an eigenvector of the linear transformation that is defined by matrix A, and the scale factor, λ, by which vector v is stretched, is called the **eigenvalue** of the linear transformation that is defined by matrix A. An eigenvector defines the characteristic direction over which the linear transformation, as defined by matrix A, preferably operates. (Number and algebra)
	(see **characteristic polynomial**)
Element (of a matrix) (or (entry of a matrix)) (AHL)	The numbers, symbols, or expressions that appear in the **rows** and **columns** of a **matrix**. For example:
	$\begin{pmatrix} 1 & 3 \\ 5 & 7 \end{pmatrix} \quad \begin{pmatrix} x \\ y \\ z \end{pmatrix} \quad \begin{pmatrix} 0 & a+b & 1 \\ ab & -4 & 0 \\ 8 & 2b^2 & 1 \end{pmatrix}$
	An element lying in row i and column j is denoted by a_{ij}. For example, the element a_{21} lies on the second row and first column, therefore in the first matrix above $a_{21} = 5$. (Number and algebra)
Element (of a set)	A member of a **set**; the objects which belong to the set. (Number and algebra)

Ellipse (of a phase portrait) (AHL)	An oval shape. An ellipse on a phase portrait shows that the system is stable. (Note that the ellipse may also take the form of a circle.) You should know that a **phase portrait** of a **dynamic system** is an ellipse when the **eigenvalues** are imaginary. (Calculus)
Empirical	This is related to something that you discover through investigation or experimentation, rather than theoretically.
	(see **empirical probability**)
Empirical probability	The practical results of an experiment, rather than theoretical results. (Statistics and probability)
	(see **experimental probability**)
Empty set	A **set** which has no **elements** in it. (Number and algebra)
Enclosed (AHL)	The area that is between a curve and a boundary, which can be an axis or another curve or line. (Calculus)
Encode (a message) (AHL)	To write a text message in an encrypted form so that it cannot be read. There are many different methods of encoding messages. Some are simple, such as Caesar's cipher, involving a shift in the alphabet. Some are more sophisticated, such as Hill's cipher, involving **matrix** algebra. Encoding a message is necessary in real-life applications, such as for internet banking, social media, military, etc. (Number and algebra)
	(see **decode (a message)**)
Enlargement (enlarge)	1 To make bigger, eg by a **scale factor** of 2, 3, etc.
	2 To make smaller, with a fractional scale factor such as 1/2 or 1/3. (Geometry and trigonometry)
Entry (of a matrix) (*plural* entries) (AHL)	(see **element (of a matrix)**)
	(Number and algebra)

E

Equal matrices (AHL)

Two **matrices** that have the same dimensions and their corresponding elements are equal. Using mathematical notation, for two matrices A and B of the same order, you can write:

$A = B$ if and only if $a_{ij} = b_{ij}$ for all i, j. For example:

$$\begin{pmatrix} x & 2 & y \\ 4 & 5 & z \end{pmatrix} = \begin{pmatrix} 0 & k & 7 \\ 4 & 5 & -2 \end{pmatrix},$$

if and only if $x = 0$, $y = 7$, $z = -2$, and $k = 2$. However,

$$\begin{pmatrix} x & 2 & y \\ 4 & 5 & z \end{pmatrix} \neq \begin{pmatrix} 0 & k & 7 \\ 3 & 4 & -2 \end{pmatrix},$$

because the corresponding elements are not equal, and

$$\begin{pmatrix} x & 2 & y \\ 4 & 5 & z \end{pmatrix} \neq \begin{pmatrix} 0 & k \\ 4 & 5 \end{pmatrix},$$

because the matrices are not of the same order. The first has dimensions 2×3 and the second 2×2. (Number and algebra)

Equality (*plural* equalities) (equal)

Where two quantities have the same value. For example, we write A = B and pronounce it as A equals B. (Number and algebra)

(see **equation**)

Equally likely (equal likelihood)

Outcomes which have the same probability. For example, we say the probability of throwing a head or a tail on a coin are equally likely. (Statistics and probability)

Equation

An expression involving an equals sign, which needs to be solved. For example, $5x + 7 = 3$. (Number and algebra)

Equation of a straight line

This can be one of three forms:

$y = mx + c$

$ax + by + d = 0$

$y - y_1 = m(x - x_1)$

(Functions)

Equation solver

An application on a **graphical display calculator** which helps you when you want to solve an equation, particularly one using exponentials.

Equidistant

At **equal** distances from. (Geometry and trigonometry)

Equilateral triangle

A triangle with three equal sides and three equal **angles**. (Geometry and trigonometry)

Equilibrium point (AHL)

A point where the state in a **dynamic system** is in equilibrium, ie the variables do not change over time. To find the equilibrium points, find the point(s) at which all the **derivatives** with respect to time are equal to zero. (Also called critical points.) (Calculus)

Equivalent (equivalency)

1. If two fractions are equivalent, the **numerator** and **denominator** have both been multiplied by the same number. For example, 2/3 is the same as 8/12 as both numerator and denominator have been multiplied by 4. (Number and algebra)

2. Equivalent **figures** have the same **area**. (Geometry and trigonometry)

Error

The difference between an amount that has been **estimated** or **rounded** and the actual value; usually expressed as a percentage. (Number and algebra)

Estimate (estimation)

To **approximate** (or to work out 'roughly') in order to see if an answer is of the right degree. May be used as a **command term**.

Euclidean geometry	Geometry (often called 'shape and space') based on the findings of the ancient Greek mathematician Euclid. It is the geometry that we are most likely to learn about in school. (Geometry and trigonometry)
Euclidean plane	A three-dimensional space according to the findings of the ancient Greek mathematician Euclid. (Geometry and trigonometry)
Euler form (of complex number) (AHL)	A complex number of the form $z = x\,\mathrm{i}y = r\,(\cos\theta + \mathrm{i}\sin\theta)$ can be also written as $z = re^{i\theta}$, ie using its **modulus** and **argument** to express it in **exponential form**. (Number and algebra)
Eulerian circuit (AHL)	A **path** that starts and ends at the same **vertex** using every edge exactly once. An Eulerian circuit can visit vertices more than once. (Geometry and trigonometry)
Eulerian graph (AHL)	A graph containing an **Eulerian circuit**. (Geometry and trigonometry)
Eulerian trail (AHL)	A **path** in a graph that visits every edge exactly once. An Eulerian trail can visit **vertices** more than once. (Geometry and trigonometry)
Euler's method (AHL)	A method used to find an approximate solution to a **first order differential equation** when it cannot be solved otherwise. Involves using the **gradient** of the **differential equation** at different points to find the curve of the approximate solution. Starting from a fixed value, x-values are chosen a fixed distance apart (**step size**) and the gradient is used to find the corresponding y-values. Plotting these points gives the curve of the approximate solution. Using a smaller step size gives a more accurate solution. (Calculus)
Evaluate (evaluation)	To work out the number, value, or amount. For example, you may be asked to evaluate $x + 2$ when $x = 7$. (Number and algebra)
Even degree (of a vertex) (AHL)	A **vertex** that has an even degree, ie it has an even number of edges joining it. (Geometry and trigonometry)
Even length (of a path)	A **path** that has an even number of edges. (Geometry and trigonometry)
Event	Something that happens, an outcome. For example, in a **probability** experiment to see how often a head is thrown when tossing a coin, you may observe one of two possible events after each toss. (Statistics and probability)
Exact	A solution which is precise. No **approximations** can be used. If you are asked to give an exact answer, you may need to use **surds**; if you give it to 2 significant figures, for example, you would not gain the mark for the answer. (Number and algebra)
Exclude (exclusion)	To leave out or not include. For example, for the integers from 1 to 5, if you exclude 5, you would be left with 1, 2, 3 and 4.
Exclusive (exclusivity)	In probability, if events are exclusive, they cannot happen at the same time. (Statistics and probability) (see **inclusive**, **non-exclusive**, **or**)
Expand (expansion)	Often used in algebra. If you expand **brackets**, it means that you multiply each term in the brackets by the expression outside the brackets. For example, when we expand the brackets in this expression: $2(x + 3y)$ it becomes: $2x + 6y$ (Number and algebra)

E

Expected frequency	How often something happens in theory. In probability, the expected number of times an outcome might be expected to occur over a given number of times. (Statistics and probability)
Expected value	The mean of a probability distribution. The expected value of the random variable X is written as $E(X)$. (Statistics and probability)
Experiment	A practical trial or test to see how many times something happens or an outcome is achieved over a number of times. For example, you could throw a coin 100 times to see how many times it lands with the head facing up. (Statistics and probability)
Experimental probability	The likelihood of an outcome that is found by doing a test or trial. Here, you are determining the probability of an event by doing a practical experiment. (Statistics and probability)
Explain	To make something clear. If you were asked to explain why the angle is 70 degrees, it would mean that you would have to give the reasons why. May be used as a **command term**.
Exponent	The exponent of a number shows how many times a number or variable is multiplied. Also known as the **power** or **index** of a number. For example: $6^3 = 6 \times 6 \times 6$ (so the exponent is 3) (Number and algebra, Functions)
Exponential decay	When a quantity decreases in a way that is proportional to its value at each stage, for example, halving at each stage. The amount by which the quantity is reduced is called the exponential decrease. The amount by which the quantity grows is called the exponential increase. (Functions)
Exponential expression	An expression with a variable that is the **exponent** of a constant. For example: 2^x (where the number 2 is raised to the **power** of x). (Number and algebra, Functions)
Exponential form (of complex number) (AHL)	(see **Euler form (of complex number)**) (Number and algebra)
Exponential function	A relationship where the variable is the exponent, or raised to a power. For example, $f(x) = 2^x$ (x is the exponent). (Functions)
Exponential growth	Where a quantity increases in a way that is proportional to its value at the current stage, for example, doubling at each stage. (Functions)
Exponential model	A real-life situation that shows growth or decay exponentially. For example, population growth. (Functions)
Exponential regression (AHL)	A **regression** curve (curve of best fit) that takes the form of an exponential function. (Statistics and probability)
Exponential relationship (AHL)	A relationship of the form $f(x) = a \cdot b^{cx+d} + e$ (defined for all real numbers), where the independent variable is in the exponent of some base b. (Functions)
Expression	A mathematical phrase which includes variables, constants, and operations. Similar to an equation but expressions do not include an **equals** (=) or **inequality** (≠) sign. (Number and algebra)
Extrapolate (extrapolation)	To predict further values, outside the range given, presuming that a trend will continue. (Functions)
Extremum (plural extrema)	The largest or smallest values of a function. For example, the maximum or minimum. (Calculus)

Face	1	The flat surface of a three-dimensional shape. For example, a cube has six square faces. (Geometry and trigonometry)
	2	Another term for **cell (on a Voronoi diagram)**. (Geometry and trigonometry)
Factor		A number which divides exactly into another number. For example, the factors of 6 are 1, 2, 3 and 6. (Number and algebra)
Factorize *or* factorise (factorization)		A way of simplifying an algebraic expression or equation, taking out anything that they have in common. For example, if we factorize $14xy - 7xy^2$, it would become $7xy(2 - y)$. (Number and algebra)
Fair		The opposite of **biased** – an experiment which happens with all circumstances being as you would expect. (Statistics and probability)
Feature		A characteristic or something which forms a significant part of a property. For example, a feature of squares is that they have four right angles.
Fibonacci numbers		A sequence of numbers where each term is found by adding the two before it. For example, the most common sequence starts with 0 and 1: $0, 1, 1, 2, 3, 5, 8, 13, 21, 34, \ldots$ This sort of sequence has many links in nature. (Number and algebra)
Figure	1	A **digit**. For example, the third figure in 134586 is 4. (Number and algebra)
	2	A diagram drawn in a text book, for example, 'See figure 6'.
Final vertex (AHL)		The **vertex** where a **walk** finishes. (Geometry and trigonometry)
Financial model		An abstract representation of a real-life situation related to money. (Number and algebra)
Find		A **command term** asking you to discover an answer and show the working. For example, you may be asked to find the largest angle in a triangle which has sides 7 cm, 8 cm, and 10 cm.
Finite		Something which stops or has an end, and can be given a value. For example: odd numbers less than 10 = a finite sequence of numbers odd numbers = an **infinite** sequence
First derivative		When you differentiate once, you will find the first derivative. If you start from y the first derivative is $\frac{dy}{dx}$. (Calculus)
First order differential equation (AHL)		A **differential equation** that uses only the **first derivative** of a function (ie not the second or any higher derivatives). For example, $\frac{dy}{dx} = \frac{5y}{3x}$ is a first order differential equation because it involves only a first derivative. (Calculus)
Fit (a model)		To move a line or curve in order to model data. (Functions)
Fit (AHL)		(see **measure of fit**) (Statistics and probability)
Fix		A **GDC** command that gives answers to a set (or fixed) number of digits. It is a command that is rarely needed, certainly for SL students.
Float (mode)		This is a function on your **GDC** that means it is not set to a certain number of decimal places.
Formula (*plural* formulae)		An algebraic equation which enables you to find something like an area, volume, etc. (Number and algebra)

F

Fractal (AHL)	A geometrical shape that includes patterns that are repeated in progressively smaller sizes, so you observe the same pattern as you zoom in but on a smaller scale. A famous example is the Koch snowflake. (Geometry and trigonometry)
Fraction	A proportion. It has a numerator and denominator, for example, ½. (Number and algebra)
Frame of reference	By changing the display dimensions on a graphic display calculator (ie by changing the frame of reference of a displayed graph), one can have access to different parts of the graph of a function. (Functions)
Free variable (AHL)	A **variable** that can take any real value. For example, in the relation $b = a - 2$, a is a free variable, as it can be any real number. But b is not free, as its value depends on a. (Number and algebra)
Frequency (*plural* frequencies)	How often something happens. Usually shown in the form of a table. (Statistics and probability)
Frequency distribution	How often each value occurs in a survey, experiment, etc. (Statistics and probability)
Frequency histogram	A vertical type of bar chart (although without gaps between the columns) to illustrate how often values appear in a distribution. (Statistics and probability)
Frequency table	A chart to present collected data in relation to how often certain results occur. (Statistics and probability)
Function	A relationship where each **element** of one **set** (for example, x) is linked with one and only one element of another set (for example, y). (Functions) (see **relationship**)
Function notation	A way of describing a function. Instead of x and y in function notation you would have x and $f(x)$ or $g(x)$. For example, we might say that a function f of x can be written $f(x) = 3x + 7$. (Functions)
Future paths (AHL)	(see **trajectory**) (Calculus)

GDC (*or graphical display calculator*)	A graphical display calculator. You can use your GDC in all papers at both SL and HL, so it is vital to practise using it from the very first day of your IB Diploma course. Your teacher will suggest which one they would like you to use.
General form (of a straight line equation)	Rather than writing an equation for a line as: $y = mx + c$ (which shows the **gradient** and *y*-**intercept**) the general form is: $ax + by + d = 0$ (where a, b and d are **integers**) For example, $y = -0.5x - 0.75$ would be rewritten as $2x + 4y + 3 = 0$ which is a better way of writing it, since everything is an integer. (Functions)
General solution (to a differential equation) (AHL)	The solution to a **differential equation** that gives all possible solutions. The general solution contains one or more **constants**. For example, the general solution to the differential equation $\frac{dy}{dx} = 4x^3$ is $y = x^4 + c$, where c is a constant. (Calculus)
Generalize (generalization)	To make a broad statement about something. (Number and algebra, Functions)
Geometric progression (*or* geometric sequence)	A set of numbers where each one is multiplied by the same number (or **common ratio**) to find the next one. Also known as a geometric sequence. (Number and algebra)
Geometric series	The sum of numbers in a **geometric sequence**. (Number and algebra)
Geometric transformation	When shapes are moved in one of four ways – **reflection**, **rotation**, **translation** or **enlargement**. (Geometry and trigonometry)
Geometry (geometric)	A branch of Mathematics involving points, lines, and shapes. (Geometry and trigonometry)
Global extremum (*plural* global extrema)	An **extremum** which is not within a given range – the largest or smallest possible. (Calculus)
Golden ratio	A relationship between numbers which is based on the **Fibonacci numbers**. It occurs often in nature. (Number and algebra)
Goodness of fit	How well data matches expected values according to certain conditions, which are stated. (Statistics and probability)
Google PageRank algorithm (AHL)	One way that Google ranks and orders webpages so that a search returns the most important page first. Pages are ranked based on the number of pages that link to them. It is an application of **graph theory** where the **vertices** represent the pages and the **edges** represent the links between pages. (Geometry and trigonometry)
Gradient	How steep something is. For example, the gradient of the line $y = 7x - 3$ is 7. (Functions)
Gradient function	Allows you to find the steepness of a line or of a curve at a certain x value. You can obtain the gradient function by **differentiation**. (Calculus)
Gradient-intercept form (of a straight line equation)	This is a way of showing the equation of a line in the form $y = mx + c$. (Functions)

G

Graph
1 A diagram which shows relationships between **variables**, often x and y. They can form straight lines, curves, etc. (Also note 'to graph' as a verb means to draw a graph or represent information in a graph. (Functions)
2 (AHL) In **graph theory**, a set of connected points. The points are called **vertices** (*singular* **vertex**). The connections are called **edges**. The edges can be given with direction (**directed graph**) or without direction (**undirected graph**). (Geometry and trigonometry)

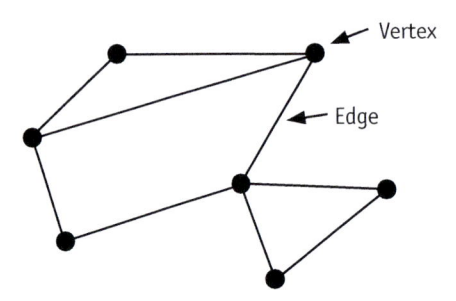

Vertex

Edge

Graph complement (AHL)
An inverse graph. The graph complement uses the same **vertices** as the graph but the vertices on the complement are only **adjacent** if they are not adjacent on the graph. (Geometry and trigonometry)

Graph theory (AHL)
The study of **graphs** where a graph is a network of connected points. Graph theory can be used to model real-world situations. You should be able to model things such as circuits and maps as graphs. It is important to note that graph theory deals with a different type of graph than the graph of an equation. (Geometry and trigonometry)

Graphing calculator
(see **GDC**)

Graphing package (*or graphing software*)
An online or electronic method of plotting graphs.

Greatest common factor (GCF) (AHL)
The largest positive number that divides exactly into two or more given numbers. For example, the GCF of $(8, 12)$ is 4, and the GCF of $(12, 18, 30)$ is 6. It is a useful tool for simplifying fractions, such as:

$$\frac{12}{30} = \frac{2 \times 6}{5 \times 6} = \frac{2}{5}$$

(Number and algebra)

Grouped data
When, instead of writing the frequency of individual items, such as number of pets, 1, 2, 3, 4, 5, 6, etc, you can write it in the form of groups of numbers, such as 1–3, 4–6, 7–9 pets, etc. (Statistics and probability)

Grouped frequency table
A chart made of **grouped data**. (Statistics and probability)

Growth factor
The **percentage** by which something increases. (Functions)

G-Solv
A graph menu on a calculator.

Half-life (AHL)	In a decaying **exponential relationship** of the form $f(t) = a \cdot b^{-t+c} + d$ $(b > 1)$, representing the decrease of x as a function of time t, the half-life, $t_{1/2}$, is defined as the time it takes for f to be halved, ie to take half of its original value $f(0) = a \cdot b^c + d$. Using **logarithms**, you can obtain a direct formula for the calculation of $t_{1/2} = -\log_b \left(\frac{1}{2} - \frac{d}{a \cdot b^c} \right)$. For $d = 0$, this simplifies to $t_{1/2} = \log_b 2$. (Functions)
Half-space	In a **Voronoi diagram**, a plane divides **3D** space into two parts, each called half-spaces. (Geometry and trigonometry)
Hamiltonian cycle (AHL)	A **path** that starts and finishes at the same **vertex** and visits each vertex exactly once. (Geometry and trigonometry)
Hamiltonian cycle of least weight (AHL)	The **path** with the smallest weight that starts and finishes at the same **vertex** and visits each vertex exactly once (a **Hamiltonian cycle**). The **travelling salesman problem** is an example of a Hamiltonian cycle of least weight. (Geometry and trigonometry)
Hamiltonian path (AHL)	A **path** in a graph that uses each **vertex** exactly once. (Geometry and trigonometry)
Handshaking lemma (AHL)	The fact that in an **undirected graph**, there will be an even number of **vertices** that have an odd degree. It is based on the idea of shaking hands at a party. If an even number of people shake hands, then each person who shakes hands will shake an odd number of other people's hands. (Geometry and trigonometry)
Hemisphere (hemispherical)	The exact half of a sphere, where a sphere is a ball shape. (Geometry and trigonometry)
Hence	Following from this. When used as a **command term** in a question, you need to use what you have already found and carry on to an answer.
Hence or otherwise	Similar to the term **hence**. When this phrase is used as a **command term** in your IB studies, it means that you may use what you have already found out in your preceding work, or that you might also gain marks for using other methods.
Highest common factor (AHL)	(see **greatest common factor**) (Number and algebra)
Histogram	A graph similar to a **bar chart**, but based on area rather than just the height of the bars. (Statistics and probability)
Homogeneous differential equation (AHL)	A **first order differential equation** that can be written as a function of $\frac{y}{x}$. That is, it can be written as: $$\frac{dy}{dx} = f\left(\frac{y}{x}\right).$$ For example: $$\frac{dy}{dx} = \frac{x^2 + 3y^2}{xy}$$ is homogeneous because it can be written as: $$\frac{dy}{dx} = \frac{1}{\frac{y}{x}} + \frac{3y}{x}, \text{ or as } \frac{dy}{dx} = \left(\frac{y}{x}\right)^{-1} + \frac{3y}{x}$$ (Calculus)

Homogeneous system of equations (AHL)	A system of two linear equations with unknown variables x and y that can take the form $$\begin{cases} a_{11}x + a_{12}y = 0 \\ a_{21}x + a_{22}y = 0 \end{cases}$$ where $a_{11}, a_{12}, a_{21}, a_{22}$ are real numbers. Equivalently, when the system is written in matrix form $$\begin{pmatrix} a_{11} & a_{12} \\ a_{21} & a_{22} \end{pmatrix} \begin{pmatrix} x \\ y \end{pmatrix} = \begin{pmatrix} 0 \\ 0 \end{pmatrix}$$ its **constant matrix** is the **zero matrix**. It accepts the zero matrix as a trivial solution. (Number and algebra)
Horizontal	Parallel to the horizon, going straight across. For example, the horizontal axis is the *x*-axis, which goes across the page. (Functions, Geometry and trigonometry) (see **vertical**)
Horizontal asymptote	A line (often drawn as dotted) which a curve approaches but never quite reaches. For example, the graph of $y = 2^x + 3$ has a horizontal asymptote of $y = 3$. (Functions) (see **vertical asymptote**)
Horizontal stretch	When a graph is distorted in the *x*-direction. If a horizontal stretch of factor k is applied, the point (x, y) is transformed to (kx, y) (Functions)
Hypotenuse	The side of a **right-angled triangle** which is opposite the right angle. It is always the longest side of the triangle. (Geometry and trigonometry)
Hypothesis (*plural* hypotheses)	A statement which tries to explain something through an educated guess or suggestion. It can be shown that it is likely to be accurate or disproved. For example, if your hypothesis is that the height of a tennis player is correlated with the speed of their serve, you could carry out a test to see if this is likely to be true. (Statistics and probability)
Hypothesis test	A statistical method to analyse observed (or experimental) data. For example, the **chi-squared test for independence** is a way of testing whether two sets of statistical data are correlated. (Statistics and probability)

i (AHL)	The **imaginary** number of unit length defined as $i = \sqrt{-1} \Leftrightarrow i^2 = -1$. Using this number, you can find the solutions of **quadratic equations** with negative discriminants (using $\Delta = -|\Delta| = i^2|\Delta|$). The solution of such equations is a **conjugate pair** of complex numbers. (Number and algebra)
IA	Internal assessment – often called coursework, which is a requirement for each of your IB Diploma subjects. In Mathematics, it is worth 20% of the final mark, both at SL and HL.
	(see **mathematical exploration**)
Identical	The same in every way.
Identify	To name or say what something is. Often used as a **command term** in your IB Diploma studies. Here, your answer must say or show what is correct from a range of different possibilities.
Identity	1 A number that keeps something the same within an operation, eg the identity for addition is 0. The identity for multiplication is 1.
	2 An equation that is true for any chosen values of the variables. For example, $2(x + 7) = 2x + 14$ is an identity as it is true for any value of x that you choose.
	(Number and algebra)
Identity matrix (AHL)	A **square matrix** where the **diagonal elements** are equal to one ($a_{ii} = 1$) and all other elements are zeros ($a_{ij} = 0, i \neq j$). For example:
	$$\begin{pmatrix} 1 & 0 \\ 0 & 1 \end{pmatrix} \quad \begin{pmatrix} 1 & 0 & 0 \\ 0 & 1 & 0 \\ 0 & 0 & 1 \end{pmatrix}$$
	The symbol I is often used to represent an identity matrix. The role of the identity matrix in matrix multiplication is the same as the role of 1 in the real numbers set. The **product** of any square matrix A, with dimensions $n \times n$, multiplied by the I matrix with the same dimensions is the original matrix A. In other words:
	$AI = IA = A$
	(Number and algebra)
Imaginary (AHL)	Any number, z, that is a multiple of i. For example, $z = k$i, is called an imaginary number. As $z^2 = -k^2$, z^2 is a real number, but z is *not* a real number. (Number and algebra)
Imaginary part (of complex number) (AHL)	The part of a complex number that is multiplied by the **imaginary unit**. For example, in a complex number $z = x + i y$, $y = \text{Im}(z)$ is the imaginary part of z. (Number and algebra)
	(see **Cartesian form (of complex number)**, **real part (of complex number)**)
Imaginary unit (i) (AHL)	(see **i**) Note that |i| = 1, thus the imaginary number i has unit length. (Number and algebra)
Imply (implication)	To suggest or indicate a conclusion can be drawn from the information given.
Impossible event	Something that cannot happen. (Statistics and probability)
In degree (of a vertex) (AHL)	The number of **edges** for which a particular **vertex** is their final vertex in a **directed graph**. (Geometry and trigonometry)
Inaccurate (inaccuracy)	Something which is not correct. (Number and algebra)

Incident (AHL)	Two objects that touch each other. In **graph theory**, a **vertex** and an **edge** that are connected to each other are said to be incident to each other. (Geometry and trigonometry)
Incident edge (AHL)	Another term for **adjacent edge**. (Geometry and trigonometry)
Incident vertex (AHL)	A **vertex** that is connected to an **edge** is said to be **incident** to that edge. (Geometry and trigonometry)
Incline	A slope. The angle of incline is measured by the number of degrees between the x-**axis** and the line. (Functions)
Included angle	An angle in a shape between two given sides. (Geometry and trigonometry) (see **non-included angle**)
Inclusive	1 Including a value, eg integers from 4 to 7 inclusive means 4, 5, 6 *and* 7. 2 In probability, if events are inclusive, they can happen at the same time. (Statistics and probability) (see **exclusive**, **non-inclusive**, **or**)
Income function	Provides the income made from selling a product or service as a function of the number of items sold, x. For example it can be given by functions like $I(x) = 2x + 30$, or $I(x) = 0.02x^2 + 5x + 20$. Also known as a revenue function. (Functions)
Increasing function	When the **gradient** of a graph is positive. (Calculus)
Indefinite integral	When you are integrating and are not given any limits between which you must integrate. The answer to an indefinite integral should always include a constant term c. (Calculus) (see **integral**)
Independence	When two variables or events are not linked. (Statistics and probability)
Independent events	When the occurrence of one event does not affect the occurrence of another. (Statistics and probability) (see **dependent events**)
Independent random variable (AHL)	An **outcome** of an experiment that is independent and random, ie the outcome does not depend on any other outcomes and the outcome could be any of the possible outcomes of the experiment. For example, the outcomes of flipping two coins are independent variables. The outcome of one coin does not depend on the outcome of the other and both coins can result in heads (H) or tails (T). (Statistics and probability)
Independent variable	A variable which doesn't depend on another variable or variables for its value. In a graph, the independent variable is usually plotted as the x coordinate. (Statistics and probability)
Indeterminate	Something which does not have a set number of answers, eg the solutions to $x + y = 7$ and $2x + 2y = 14$.
Index (*plural* indices)	The **power** to which a number is raised, written as the small number to the upper right. For example, for 2^3 (or 2 cubed), the index is 3. (Number and algebra) (see **exponent**)
Index laws (*or* index rules)	Rules for using indices, for example: $x^3 \times x^4 = x^7$ (Number and algebra) (see **laws of exponents**)

Index notation	A way of writing numbers using the **powers** of those numbers. For example: $2 \times 2 \times 2 \times 3 \times 3 \times 5 \times 5 = 2^3 \times 3^2 \times 5^2$ (Number and algebra)
Inequality (*plural* inequalities)	1 Where two values are not equal ($\neq$). 2 A mathematical statement comparing two numbers or expressions using inequality signs, $<, >, \leq, \geq$. (Number and algebra)
Infer (inference)	When a conclusion can be drawn from what has been worked out.
Infinite geometric series (AHL)	The sum of an infinite **geometric sequence**. When the **partial sum of a geometric series**, S_n, is considered as $n \to \infty$, the resulting sum with infinite terms, S_∞, will **converge** to a finite value if, and only if, the corresponding geometric sequence has decreasing terms, ie $|r| < 1$. Then $S_\infty = \frac{u_1}{1-r}$. (Number and algebra)
Infinite set	A set which does not ever end, eg the set of prime numbers.
Infinitesimal (infinitesimally)	So small we cannot measure it; extremely close to zero (but not zero). (Calculus)
Infinity	A very large number that cannot be classified.
Inflation	Price increase. (Where prices fall, we call it deflation.) (Number and algebra)
Inflation rate	The amount by which prices rise, usually expressed in the form of a percentage. (Number and algebra)
Inflexion, point of (AHL)	(see **point of inflexion**) (Calculus)
Initial conditions	What happens at the start of an event, eg a cake may start at room temperature and then be put into an oven. The initial condition is room temperature. (Functions)
Initial point (AHL)	The starting point when a vector is drawn as a **line segment**. (Geometry and trigonometry) (see **terminal point**)
Initial state matrix (or initial state probability matrix) (AHL)	A way of displaying the probability of starting in each state of a **Markov chain**. For example, if there is a 40% chance of starting in state A and a 60% chance of starting in state B, the initial state matrix can be written as shown, where s_0 means the initial state: $s_0 = \begin{pmatrix} 0.6 \\ 0.4 \end{pmatrix}$ (Statistics and probability)
Initial value	A starting amount. For example, we refer to the initial value of an **exponential function**. (Functions)
Initial value problem (AHL)	A **differential equation** problem that has an initial condition. For example, solve: $\frac{dy}{dx} = \sqrt{\frac{x}{y}}$, given that the curve passes through the point (3,1). (Calculus)
Initial vertex (AHL)	The **vertex** from which a **walk** starts. (Geometry and trigonometry)
Inner product (AHL)	(see **scalar product**) (Geometry and trigonometry)
Input (value)	An amount which is substituted into an expression. (Functions) (see **output**, **relationship**)

I

Inquirer	The IB learner profile suggests ten qualities that you should be aiming for. One of these is to be an 'inquirer' which means being curious about what is going on around you. In your mathematics studies, this may mean that when you are taught a new method you investigate it further to see where it came from or where it leads.
Instantaneous	When something such as acceleration or velocity is calculated at a certain time (or instant in time). (Calculus)
Integer	A whole (counting) number or its negative. (So fractions, decimals, etc cannot be referred to as integers.) (Number and algebra)
Integral	1 Related to **integer**, where there are no fractional parts, only whole numbers. 2 The space that is found underneath the graph of an equation. (Calculus) (see **definite integral**, **indefinite integral**)
Integral calculus	A form of mathematics where **differentiation** and **integration** are used. (Calculus)
Integral sign	The symbol which represents integration ($\int$). (Calculus)
Integrate (integration)	The reverse process of **differentiation**. This can be used as a **command term**. (Calculus)
Integration by inspection (AHL)	When **integration** can be done in an obvious way because the **antiderivative** is clear. Integration by inspection is used when the integration requires no further techniques. The formula book contains a list of standard integrals. (Calculus)
Integration by substitution (AHL)	A technique for **integration** when the function is in the form $f(g(x)) \times g'(x)$. Let $u = g(x)$, then $du = g'(x)dx$, then $\int f(g(x)) \times g'(x) = \int f(u)du$ For example, $\int \sin(2x+3)\,dx$ Let $u = 2x + 3$, then $du = 2\,dx$ (so $dx = \frac{1}{2}du$) Then, $\int \sin(2x+3)\,dx = \int \sin(u) \times \frac{1}{2}du$ $= \frac{1}{2}\int \sin(u)\,du = -\frac{1}{2}\cos(u) + c = -\frac{1}{2}\cos(2x+3) + c$ (Calculus)
Intercept	Where a graph cuts one of the axes – referred to as the **x-intercept** or **y-intercept**. (Functions)
Interest	The amount paid when money is used for another purpose. For example, when money is invested in a savings account, the bank pays you an extra amount (ie the interest); or when you borrow money, you pay a charge (or interest) on the money borrowed. (Number and algebra)
Interest rate	Used to calculate how much is paid for borrowing money (or how much the bank pays you to invest money in a savings account), expressed as a percentage. (Number and algebra)

Internal assessment (IA)	Often called coursework, this is a piece of work you do which is an assessment requirement for each of your IB Diploma subjects. In Mathematics, it is worth 20% of the final mark, both at SL and HL. (see **mathematical exploration**)
International-mindedness	The IB learner profile suggests ten qualities that you should be aiming for. One of these is to have 'international-mindedness' which means understanding how other countries and cultures work. In Mathematics, this can mean that you learn different methods of working out from people from other countries and the different notation that the IB will accept, according to the traditions of different countries.
Interpolate (interpolation)	To estimate a value (of a function) within a known range of given values. Similar to **extrapolate** but the range of values is known. (Functions)
Interpret (interpretation)	To work out how the results of something can be shown in real life. Can be used as a **command term**.
Interquartile range (IQR)	The value of the **lower quartile** taken from the value of the **upper quartile**. (Statistics and probability)
Intersection (of curves or lines)	Where lines or graphs cross. (Functions)
Intersection (of sets)	The elements that two or more **sets** have in common. On a **Venn diagram**, it is where the circles overlap. (Number and algebra)
Interval	Defines the set of numbers between two points or within a range of values. (Number and algebra, Calculus)
Invalid	Something which cannot be true or allowed. For example when solving a quadratic equation to find the length of a line, you must reject a negative answer. This will be invalid because the length of a line must be > 0.
Invariant (AHL)	The property that an object remains unchanged after a transformation is applied. For example, the Cartesian axes are invariant to **stretches**. This means that points on the x-axis, with coordinates $(x, 0)$, do not change under a vertical stretch of factor q, as they are transformed to $(x, 0 \times q)$, ie $(x, 0)$; similarly, points on the y-axis, with coordinates $(0, y)$, do not change under a **horizontal stretch** of factor p, as they are transformed to $(0 \times p, y)$. As another example, all objects are invariant to rotations of $360°$. (Functions)
Invariant state (AHL)	Points (x, y) that do not change position after the application of a linear transformation (eg in terms of matrix A), are invariant with respect to the transformation. These are all points that satisfy the matrix equation $A \begin{pmatrix} x \\ y \end{pmatrix} = \begin{pmatrix} x \\ y \end{pmatrix}$, which produces the equation of the straight line containing all invariant points under the transformation, or its invariant line. (Number and algebra)
Inverse (of a matrix) (AHL)	The matrix A^{-1} which has product I when multiplied by a matrix A in any order: $AA^{-1} = A^{-1}A = I$ A^{-1} is the inverse of matrix A and I is the **identity matrix**. This is a special case in which matrix multiplication becomes **commutative**. Note that -1 is not an exponent but only denotes the matrix inversion. In your IB Diploma studies at HL, finding the inverse of only 2×2 matrices will be examined, and in the case of matrix $A = \begin{pmatrix} a & b \\ c & d \end{pmatrix}$, the inverse is calculated as $A^{-1} = \frac{1}{|A|} \begin{pmatrix} d & -b \\ -c & a \end{pmatrix}$, where $|A|$ is the **determinant** of matrix A. (Number and algebra)

I

Inverse function — A **function** which reverses the effect of the another function. For example, the inverse function of addition is subtraction, whilst the inverse of multiplication is division. (Functions)

Inverse normal calculation — Used to find statistical values when working on problems involving the **normal distribution** in which you are given the area; you need to use your **GDC** to do these calculations. (Statistics and probability)

Inverse proportion (*or* **inverse variation**) — When two **variables** are linked in a way that as one variable (x) increases, the other (y) decreases in **proportion**. (Functions)
(see **direct proportion**)

Invertible matrix (AHL) — A matrix A that has an inverse A^{-1} such that $AA^{-1} = A^{-1}A = I$. (Number and algebra)

Investigate (investigation) — To look in depth at something, to examine thoroughly, to try and work out the truth as to why something happens. (Often used as another name for the **IA** in Mathematics at IB or as a **command term**.)

Investment — An amount which is put into a bank or savings account in order to raise more money. (Number and algebra)

invNorm — A button on your calculator which is used for working with the **normal distribution**. (Statistics and probability)

IQR (Interquartile range) — (see **interquartile range**) (Statistics and probability)

Irrational — A number which cannot be written as a fraction. (Here, irrational means 'not **rational**'.) The most common examples are π, $\sqrt{2}$ and $\sqrt{3}$.
(Number and algebra)

Irreducible Markov chain (AHL) — When all states in a **Markov chain** are connected with each other. (Statistics and probability)

Irregular — Usually used to describe shapes; an irregular shape is one where its sides are not all of the same length.

Isocline (AHL) — A set of points in a **slope field** that have the same gradient, ie:
$$\frac{dy}{dx} = k$$
where k is some **constant**. (Calculus)

Isomorphic (isomorphism) (AHL) — When a **graph** is in a different form but still has the same number of **vertices** and **edges**, and the connectivity is the same. For example, the two graphs below are isomorphic because although they look different, the vertices and edges are the same. A is connected to B and D and A′ is connected to B′ and D′ and so on. (Geometry and trigonometry)

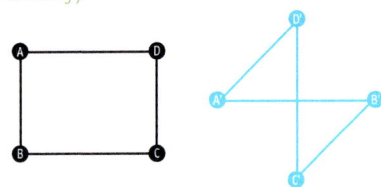

Isomorphic invariants (AHL) — The parts of a graph that do not change in an **isomorphic** graph. The number of **edges** and **vertices** are isomorphic invariants because they do not change under an isomorphism. (Geometry and trigonometry)

Iterative technique (AHL) — To repeat a set of steps where the outputs from the previous step are the inputs to the next one. For example, in order to generate **fractals**, you start with a segment and divide it, redraw it, then repeat the same process on the redrawn image over and over again. (Geometry and trigonometry)

Justify (justification)	To show or prove that something is true. Often used in relation to the **IA**, where you need to justify a solution from your data.
Kilo-	Used as a prefix to denote multiplication by 1000. Most often used in weights and measures, eg 1000 grams = 1 kilogram. (Number and algebra)
Kinematics (AHL)	The Mathematics of the movement of objects. For example, finding the **acceleration** of an object. (Geometry and trigonometry, Calculus)
Kite	A four-sided flat shape which has two pairs of adjacent equal sides. (Geometry and trigonometry)
Knowledgeable	The IB learner profile suggests ten qualities that you should be aiming for. One of these is to be 'knowledgeable' which means understanding and being interested in many different areas. In your own study of Mathematics, this could relate to how Mathematics can be used in real-life situations.
Kruskal's algorithm (AHL)	An **algorithm** used to find a **minimum spanning tree**. Pick the edges in order from smallest to largest weight ensuring that no cycles are created. Keep picking edges until all vertices are in the same tree. (Geometry and trigonometry)
k-step walk	The number of edges followed in a walk from one vertex to another. For example, if a walk goes through 4 edges to get from vertex A to B, it is a 4-step walk. (Geometry and trigonometry)

L

Label	A **command term** often used at IB Diploma level which means to put information onto a diagram (eg you might be asked to write the side lengths and angles on a diagram).
Laws of exponents *or* laws of indices	A set of rules for working out sums including **powers**, for example, when you multiply **indices** together, you add the powers. For example, $a^5 \times a^7 = a^{5+7} = a^{12}$. (Remember **exponents** are also called powers or indices.) (Number and algebra)
Laws of logarithms (AHL)	Algebraic expressions that translate the **laws of indices** in logarithmic notation. Specifically, • $\log_c(a \times b) = \log_c a + \log_c b$ • $\log_c\left(\dfrac{a}{b}\right) = \log_c a - \log_c b$ • $\log_c a^n = n \log_c a$ In your IB Diploma exams at HL, the base of the logarithm will be either 10 or e. (Number and algebra)
Leading diagonal (of a matrix)	The line which goes from top left to bottom right, whatever size the matrix is. (Number and algebra)
Least common multiple (LCM) (AHL)	The smallest positive number that is a **multiple** of all numbers given. For example, the LCM of (8,12) is 24; the LCM of (2,3,5) is 30. (Number and algebra)
Least squares regression curve (*or* least squares regression) (AHL)	The curve (or line for **linear regression**) that best fits the points on a **scatter diagram**. It is the curve that has the smallest sum of the squares of the differences in the y-values of the curve (or line) and of the points in the scatter graph (ie the smallest **sum of square residuals**). This is usually done using technology. The formulae for finding the equation of the regression lines of x on y and of y on x are given in the formula book. In the exam, you will need to be familiar with linear, quadratic, cubic, exponential, power, and sine least squares regression curves on your calculator. (Statistics and probability)
Leibniz's notation (*or* Leibniz notation)	The way of writing differentiation as $\dfrac{dy}{dx}$ where dx and dy represent infinitesimally small increments of y and x. (Calculus)
Level of significance	Another term for **significance level**. (Statistics and probability)
Limit	A value that is approached but never quite reached by a function. Used when you are stating **asymptotes** on graphs. (Calculus) (see **limit notation**)
Limit notation	A way of expressing that the value of a function as its input approaches a value. The notation $\lim_{x \to a} f(x)$ means the limit of the function $f(x)$ as x tends to a. (Calculus)
Line	A one-dimensional figure which joins two points in the most direct way possible and continues indefinitely. (Geometry and trigonometry)
Line graph	A diagrammatical representation of a data set where all points are joined together by straight lines. (Statistics and probability)
Line of best fit (*or* best fit line)	Points on a graph, such as a **scatter diagram**, joined up so that they are as close as possible to a straight line to represent the trend of the values. The line should go through as many of the points as possible but must still be a straight line. (Statistics and probability)
Line of regression	Used in statistics for showing the **gradient** and **intercept** of the line which best fits the data. (Statistics and probability)

Not to be photocopied

Line of symmetry	1. Something which divides a graph in two matching halves. (Functions)
	2. Something which divides a shape in half so that each half perfectly match. A circle has infinite lines of symmetry, but a triangle may have 1, 3, or no lines of symmetry. (Geometry and trigonometry)
	(see **axis of symmetry**)
Line segment	A line which joins two endpoints in the most direct way possible. (Geometry and trigonometry)
Linear (linearity)	Forming a straight line. (Functions)
Linear combination (of random variables) (AHL)	A sum (or difference) of **random variables** where the random variables are not multiplied by each other. The random variables may be multiplied by **constants**. The **expected value** $E(X)$ and variance $Var(X)$ can be combined using the **formulae**:
	$E(a_1 X_1 \pm a_2 X_2) = a_1 E(X_1) \pm a_2 E(X_2)$
	$Var(a_1 X_1 \pm a_2 X_2) = (a_1)^2 Var(X_1) + (a_2)^2 Var(X_2)$
	These formulae are given in the formula book. You need to know that a linear combination of normal random variables will follow a **normal distribution**:
	$X \sim N(\mu, \sigma^2) \Longrightarrow \bar{X} \sim N\left(\mu, \frac{\sigma^2}{n}\right)$
	(Statistics and probability)
Linear combination (of vectors) (AHL)	Adding and subtracting scalar multiples of **vectors**. For example, given $v = 2u + 3w$, $2u$ is a scalar multiple of u, $3w$ is a scalar multiple of w, so v is a linear combination of u and w. (Geometry and trigonometry)
Linear correlation	The extent to which a relationship between two variables is linear. (Statistics and probability)
Linear equation	An expression involving an equals sign, which needs to be solved, and which would form a straight line when graphed. For example: $5x + 7 = 3y$ (Number and algebra, Functions)
Linear expression	As linear equation, but it is not equal to anything. (Number and algebra, Functions)
Linear function	A function which forms a straight line. (Functions)
Linear graph	A graph which forms a straight line when plotted. (Functions)
Linear inequality	An expression involving an inequality sign, such as <, >, etc and which needs to be solved, for example:
	$5x + 7 < 3y$
	It would form a region when graphed. (Number and algebra)
Linear model	A representation of something in real life by a straight line model, eg the line $y = 5x$ can be used to represent a hill which goes from A to B and which has a **gradient** of 5. (Functions)
Linear motion (AHL)	Movement in a straight line. (Geometry and trigonometry)
Linear regression	A mathematical means of finding the **line of best fit** on a **scatter diagram**. (Statistics and probability)
	(see **least squares regression curve**)

L

Linear relationship	A relationship between two variables that means that when you plot them against each other, they form a straight line. (Statistics and probability)
Linear transformation (of a single random variable) (AHL)	Changing a **random variable** by multiplying or dividing it by a constant and/or adding or subtracting a constant. Given a random variable X, it can be transformed into a random variable Y so that $Y = aX + b$, for some constants a and b: $$E(Y) = E(aX + b) = aE(X) + b$$ $$\text{Var}(Y) = \text{Var}(aX + b) = a^2\text{Var}(X)$$ (Statistics and probability)
Linearize (linearization) (AHL)	When a **best-fit straight line** ($y = ax + b$) is drawn on a **log-log graph**, where x and y are the logarithms of X and Y respectively, then one can use **laws of logarithms** to reveal the nature of the relationship of X and Y, ie the original quantities before logarithmic **scaling** is applied. Specifically, $$\log Y = a \log X + b \implies \log Y = \log(10^b X^a) \implies Y = 10^b X^a$$ so a **power relationship** is revealed between X and Y. When a best-fit straight line ($y = aX + b$) is drawn on a semi-log graph, where y is the logarithm of Y, then one can use laws of logarithms to reveal the nature of the relationship of X and Y, ie the original quantities before logarithmic scaling is applied. Specifically, $$\log Y = aX + b \implies Y = 10^b 10^{aX}$$, so an **exponential relationship** is revealed between X and Y. Using the best-fit straight line to obtain a power or an exponential relationship for the variables before scaling is applied is a process called linearization. (Functions)
List	A frequent **command term** which asks you to write down a series of numbers or words.
Loan	An amount of money lent (for example, by a bank) which will be paid back over time, usually with **interest**. (Number and algebra)
Local maximum (*plural* local maxima)	A point where a graph changes its direction – from positive gradient, through 0 to a negative one. It may not be the highest point on the whole graph. The gradient at the local maximum is 0. (Calculus)
Local minimum (*plural* local minima)	A point where a graph changes its direction – from negative gradient, through 0 to a positive one. It may not be the lowest point on the whole graph. The gradient at the local minimum is 0. (Calculus)
Logarithm	The **index** to which a base number must be raised in order to make a given number, for example, the logarithm of 1000 to base 10 is 3. (Number and algebra)

Logistic function (AHL)	A function of the form $f(x) = \frac{L}{1 + Ce^{-kx}}$, where $C, L, k > 0$. Because of its graph's 'S' shape, it is often also called a sigmoid function. The logistic function is used to describe the growth of a population that has **restrictions on** its **growth**, ie restrictions on its maximum size, most typically restrictions set by the availability of resources and/or space. The logistic function grows almost exponentially at first, with its graph being **concave up**, but its growth is progressively slowing down, first presenting a **point of inflexion** (also known as the sigmoid point), after which the function continues growing, with its graph now being **concave down**, and finally approaching the value $f(x) = L$. This is the maximum value of the population that has been set by the restrictions on growth that have been set for the population that is under study. The parameter L is called the **carrying capacity** of the population or the logistic function. The parameter k relates to the rate with which the logistic function grows: the larger the value of k, the faster the growth. The parameter C relates to the initial value of the population and the carrying capacity, L, through the formula $C = \frac{L}{f(0)} - 1$. (Functions)
Logistic model (AHL)	In problems involving population growth, and modelling data points positioned on the **Cartesian plane** in a form that resembles the (sigmoid) shape of the graph of a **logistic function**, one may use technology to obtain the best-fit values for the logistic function parameters C, L, and k. (Functions)
Log-log graph (AHL)	In an analysis of **bivariate data** where values of both variables extend over several orders of magnitude, one can use **scaling** to obtain the logarithms of both quantities, which can then be plotted on what is known as a log-log graph. (Functions)
Long-term probability (AHL)	The **probability** of an event or state measured over a long period of time. (Statistics and probability)
Loop (AHL)	An **edge** that connects a **vertex** to itself. (Geometry and trigonometry)
Loss	When you or a business gets back (or earns) less money than you paid (or was invested), then there is a loss. (Number and algebra)
Lower bound	1. The smallest value that fits a set of criteria. For example, when looking at 500 to the nearest 100, the lower bound is 450. (Number and algebra) 2. The smallest value a measurement can take. For example, in the **travelling salesman problem**, the lower bound is the smaller number in the range of lengths of the minimum route between cities. (Geometry and trigonometry) (see **upper bound**)
Lower boundary	Calculation of halfway between the top of one **class** and the bottom of the class above it. For example, for 5 – 8 and 9 – 12 the lower boundary of the 9 – 12 class is 8.5. (Statistics and probability) (see **upper boundary**)
Lower quartile	Where a data set is divided up into quarters (or **quartiles**), the lower quartile is the number which is a quarter of the way up from the lowest point in the data. This can either be calculated from a list of numbers or shown on a **cumulative frequency graph**. (Statistics and probability) (see **upper quartile**)
Lowest common multiple (AHL)	(see **least common multiple**) (Number and algebra)

M

Magnitude (of a vector) (AHL)
The size of a vector. Given a vector v, it is denoted $|v|$. It is defined as the **square root** of the sum of the square of each **component of the vector**. For example, the magnitude of the vector $v = -3i + 4j + k$ is $|v| = \sqrt{(-3)^2 + 4^2 + 1^2} = \sqrt{26}$.
(Geometry and trigonometry).

Main diagonal (of a matrix) (AHL)
(see **diagonal (of a matrix)**) (Number and algebra)

Manipulate (manipulation)
A word which asks you to change an **expression** so that it becomes simpler. For example if $y = 2x - 5 + 7x + (7/2)$ it can be manipulated to $y = 9x - 1.5$

Mantissa
When a value is written in **standard form**, the mantissa is the number that appears before the multiplication sign. For example, in standard form $2563 = 2.563 \times 10^3$, and the mantissa is 2.563. (Number and algebra)

Many to one function
When more than one of the x (**domain**) values maps to the same y (**range**) value. (Functions)

Mapping
To change something by applying a rule, eg the set of whole numbers can be mapped to the set of square numbers by applying the rule that each number maps to its square. (Functions)

Mapping diagram
A visual way to show how x values map to the equivalent y values. (Functions)

Markov chain (AHL)
A sequence of states where the **probability** of the next state is only dependent on the current state. For example, if it is a sunny day today, there is an 85% chance it will be sunny tomorrow and a 15% chance it will be rainy. If it is a rainy day today, there is a 20% chance it will be sunny tomorrow and an 80% chance it will be rainy. It does not matter whether it was sunny or rainy yesterday or the day before. The states Sunny and Rainy form a Markov chain as displayed in the **transition diagram** below. (Statistics and probability)

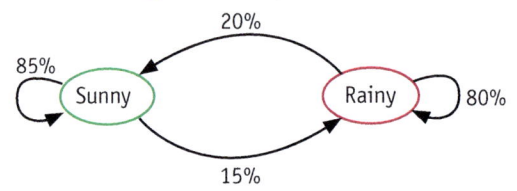

Markov property (AHL)
The property in a **Markov chain** that the next state is dependent only on the current state and not any others that may have come before. (Statistics and probability)

(see **memory-less**)

Matched pairs (AHL)
Putting experimental data together into groups of two where the two data points are as similar as possible. Using matched pairs can reduce the variability of the **sample**. For example, to evaluate the effect of a new revision guide for IB Diploma Mathematics, you could look at exam results of a group of students and pair up two students who had similar mock exam scores, one who used the new guide and one who did not. (Statistics and probability)

Mathematical exploration
Used in the **internal assessment** for IB Diploma Mathematics, this is a piece of written work which gives you an opportunity to investigate an aspect of Mathematics which interests you. It counts for 20% and is marked by your teacher but moderated by the IBO. In order to do well, you need to make sure that all your work is set out well and explained clearly using appropriate language.

Mathematical model — Something used to try and reproduce real-life situations using Mathematics. A model might help us understand or predict what happens when prices, **interest rates**, or temperatures change, for example. (Functions)

Matrix (*plural* matrices) (AHL) — A collection of numbers, symbols, and/or expressions arranged in **rows** and **columns** in a rectangular block. For example:

$$\begin{pmatrix} 3 & 4 \\ 2 & 1 \end{pmatrix} \quad \begin{pmatrix} 6 & 4 & 9 \\ 5 & 7 & 2 \\ 1 & 3 & 8 \end{pmatrix} \quad \begin{pmatrix} a & b \\ c & 2x \\ 9 & 0 \end{pmatrix} \quad \begin{pmatrix} x+1 \\ y^2 \end{pmatrix}$$

In general, a matrix A with m rows and n columns is denoted by:

$$A = \begin{pmatrix} a_{11} & a_{12} & \cdots & a_{1n} \\ a_{21} & a_{22} & \cdots & a_{2n} \\ \vdots & \vdots & \ddots & \vdots \\ a_{m1} & a_{m2} & \cdots & a_{mn} \end{pmatrix}$$

Matrices are often used to store information, eg this table shows the choice of sport per gender for the members of a sports club.

	Tennis	Football	Swimming
Male	45	67	46
Female	56	34	38

This information may also be written in matrix form:

$$\begin{pmatrix} 45 & 67 & 46 \\ 56 & 34 & 38 \end{pmatrix}$$

where each row represents a gender and each column a different sport. (Number and algebra)

Matrix equation (AHL) — An equation involving **matrices**, in which the unknown variable is a matrix. For example:

$$\begin{pmatrix} 1 & 0 \\ 0 & -1 \end{pmatrix} + \begin{pmatrix} x & y \\ z & w \end{pmatrix} = \begin{pmatrix} -5 & 1 \\ 2 & -1 \end{pmatrix}$$

or

$$A + X = B$$

where X is the variable matrix. Matrix equations are often used when solving **simultaneous equations**. For example, a system of linear equations with unknowns x and y, such as

$$\begin{cases} x + 2y = 5 \\ 3x - 8y = 12 \end{cases}$$

can be represented by the matrix equation:

$$\begin{pmatrix} 1 & 2 \\ 3 & -8 \end{pmatrix} \begin{pmatrix} x \\ y \end{pmatrix} = \begin{pmatrix} 5 \\ 12 \end{pmatrix}$$

or

$$AX = B$$

where

$$A = \begin{pmatrix} 1 & 2 \\ 3 & -8 \end{pmatrix}, \ B = \begin{pmatrix} 5 \\ 12 \end{pmatrix}, \text{ and } X = \begin{pmatrix} x \\ y \end{pmatrix}$$

(Number and algebra)

(see **equal matrices, variable matrix**)

Matrix multiplication (AHL)		The operation of multiplying a **matrix** by another matrix, producing a third matrix.
		Two matrices can be multiplied only if the number of **columns** of the first matrix is equal to the number of **rows** of the second. In general, the result of the multiplication of a matrix A (m × n) by a matrix B (n × k) is a third matrix C with **dimensions** m × k:
		$AB = C$ or $A \cdot B = C$
		The operation involves pairing each row of the first matrix with the corresponding column of the second. For example:
		$$\begin{pmatrix} 0 & 6 \\ -4 & 7 \\ 3 & 5 \end{pmatrix} \begin{pmatrix} 1 \\ 2 \end{pmatrix} = \begin{pmatrix} 0 \times 1 + 6 \times 2 \\ -4 \times 1 + 7 \times 2 \\ 3 \times 1 + 5 \times 2 \end{pmatrix} = \begin{pmatrix} 12 \\ 10 \\ 13 \end{pmatrix}$$
		3 × 2 2 × 1 3 × 1
		(Number and algebra)
Matrix of coefficients (AHL)		(see **coefficient matrix**) *(Number and algebra)*
Matrix transformation (AHL)		Using **matrix multiplication** to transform points on a graph. To transform a point (x,y), write it as a point matrix $\begin{pmatrix} x \\ y \end{pmatrix}$. To transform that point by a transformation matrix $\begin{pmatrix} a & b \\ c & d \end{pmatrix}$, multiply the transformation matrix by the point matrix to get the point in the image, ie find
		$$\begin{pmatrix} a & b \\ c & d \end{pmatrix} \times \begin{pmatrix} x \\ y \end{pmatrix}$$
		(Geometry and trigonometry)
Maximize (*or* maximise)		To make a value as large as possible. *(Calculus)*
		(see **minimize**)
Maximum (*plural* maxima)		A point where a graph changes its direction – from a positive gradient through 0 to a negative one. The gradient at a maximum is 0. *(Functions, Calculus)*
Maximum point (AHL)		The highest point on a curve in a range of values. A maximum point can be 'local maximum', ie a maximum point in a subset of **values**, or a 'global maximum' ie the maximum point of the entire function. *(Calculus)*
		(see **minimum point**)
Mean		An **average** which you work out by adding all of the numbers you have and then dividing by how many there are. *(Statistics and probability)*
Mean point		Something plotted on a **scatter diagram** in order to help draw a **line of best fit** (this involves the **averages** of both sets of data). *(Statistics and probability)*
Measure of central tendency		Central tendency relates to the likelihood that values of a random variable will cluster around the **mean**, **median**, or **mode**. When we refer to the measure of central tendency, these are most often called averages. *(Statistics and probability)*
Measure of dispersion		In statistics, how far individual values are from the **mean** of the group. *(Statistics and probability)*
Measure of fit (AHL)		A measure of how closely the **model** reflects the data. *(Statistics and probability)*
Measurement error (*or* measuring error)		The difference between what something really measures and what someone has measured it as. *(Number and algebra)*

Measurement limitation	Since measuring devices cannot provide infinite precision, eg rulers do not offer marks below the mm level, you should always be aware of the measurement limitations for estimating the precision of measured quantities. In the example of rulers, the precision in measuring lengths is at the mm level. (Number and algebra)
Measurement system	A way of using units of measurement. We tend to use the decimal system of measuring, involving metres (m), kilometres (km), etc., but there are also other systems such as **degrees** or **radians** for measuring **angles** (radians are more likely to be used at AHL). (Number and algebra)
Median	The middle number in a sequence of numbers, when they are put in order. (Statistics and probability)
Memory-less (AHL)	A **probability distribution** is memory-less when the **probability** of an event is not dependent on the outcome of any previous events. For example, each time a **die** is rolled the probability of the outcome is not affected by any previous rolls of the die, so it is memory-less. (Statistics and probability)
Menu	This is GDC notation and represents the series of choices on a GDC from which you pick the one you need.
Mid-interval values	When you have **grouped data**, this is the middle number. For example, for the group 7–11, the mid-interval value is 9. (Statistics and probability)
Mid-point (of a line segment)	The middle of, or halfway along, a line which joins two points. (Geometry and trigonometry)
Milli-	A prefix which represents a thousandth (or 1/1000), eg a millimetre is a thousandth of a metre. (Number and algebra)
Million	The number 1,000,000 which represents one thousand thousands. It can be written as 10^6. (Number and algebra)
Millionth	The number 1/1,000,000. It can be written as 10^{-6} or 0.000,001. (Number and algebra)
Minimize (*or* minimise)	To make a value as small as possible. (Calculus) (see **maximize**)
Minimum (*plural* minima)	A point where a graph changes its direction – from negative gradient through 0 to a positive one. The gradient at a minimum is 0. (Functions, Calculus)
Minimum point (AHL)	The lowest point on a curve in a range of **values**. A minimum point can be 'local minimum', ie it is a minimum point in a subset of values, or a 'global minimum' ie the minimum point of the entire function. (Calculus) (see **maximum point**)
Minimum spanning tree (MST) (AHL)	A **spanning tree** of a graph that has the smallest possible total weight of the edges. (Geometry and trigonometry)
Modal class	When you have a set of values and put them in groups (or **classes**), the class which contains the highest number of values. (Statistics and probability)
Mode	The number which appears the most amount of times. For example, for 2, 3, 8, 6, 2, 5, 2, 2 the mode is 2. (Statistics and probability)
Model (modelling)	Using Mathematics to solve real-life problems. (Number and algebra, Functions) (see **mathematical model**)
Modulus	The size of a number regardless of whether it is positive or negative. The modulus of a number will always be positive. For example, the modulus of –30 is 30. (Also known as **absolute value**.) (Number and algebra)

M

Modulus (of complex number) (*plural* moduli) (AHL)
A complex number $z = x + iy$, can be represented on the **complex plane** with a position vector $\begin{pmatrix} x \\ y \end{pmatrix}$ having magnitude $r = \sqrt{x^2 + y^2}$. This is the complex number's modulus. (Number and algebra)

Modulus-argument form (of complex number) (AHL)
A complex number $z = x + iy$ can be also written as $z = r(\cos\theta + i\sin\theta)$, ie using its **modulus** and **argument**, or its polar coordinates (r, θ). (Number and algebra)

(see **polar form (of complex number)**)

Monotonic relationship
A relationship between two variables such that as one variable increases, the other variable either only increases or only decreases. (Statistics and probability)

Multigraph (AHL)
A graph in which there can be more than one **edge** between the same two **vertices**. (Geometry and trigonometry)

Multiple
The multiples of a figure are numbers which are 1 ×, 2 ×, 3 ×, etc that figure. For example, multiples of 6 include 6, 12, 18, 24, and 30.
(Number and algebra)

Multiplication (multiply)
A mathematical operation where you repeatedly add to work out how many are in certain numbers of specified groups. For example, if you have 10 boxes of chocolates, each containing 7 chocolates, the total number of chocolates is 10 multiplied by 7 (or 10 × 7) which is 70. (Number and algebra)

Multiplicative identity (AHL)
The property stating that when any real number is multiplied by 1, the product is the original number: $a \times 1 = 1 \times a = a$, for all real numbers. The number 1 is called the multiplicative identity element. The multiplicative identity property is extended as well to other sets, such as complex numbers, where the identity element is also 0, and matrices, where the identity element is the **identity matrix** I. (Number and algebra, Geometry and trigonometry)

Multiplicative inverse (AHL)
The matrix A^{-1} that when multiplied by a matrix A, in any order, results in the **identity matrix**.

$$AA^{-1} = A^{-1}A = I$$

Then A^{-1} is the multiplicative inverse of A. Note here that the -1 in A^{-1} does not imply an exponent.

For example $\begin{pmatrix} 7 & 5 \\ 4 & 3 \end{pmatrix}$ is the multiplicative inverse of $\begin{pmatrix} 3 & -5 \\ -4 & 7 \end{pmatrix}$, because

$$\begin{pmatrix} 7 & 5 \\ 4 & 3 \end{pmatrix}\begin{pmatrix} 3 & -5 \\ -4 & 7 \end{pmatrix} = \begin{pmatrix} 3 & -5 \\ -4 & 7 \end{pmatrix}\begin{pmatrix} 7 & 5 \\ 4 & 3 \end{pmatrix} = \begin{pmatrix} 1 & 0 \\ 0 & 1 \end{pmatrix}$$

The same concept can be viewed in real numbers, where two numbers are multiplied to give the multiplicative identity 1:

$$a \times \frac{1}{a} = 1, a \neq 0$$

All real numbers have a multiplicative inverse, except for zero. For example, -5 is the multiplicative inverse of $-\frac{1}{5}$, because $-5 \times \left(-\frac{1}{5}\right) = 1$.
For numbers, we may also use the term **reciprocal**. (Number and algebra)

Multiplicative probability law
This is a formula used in probability to find the probability of the intersection of two sets. You will use this law in your IB Diploma Mathematics course, but you don't need to remember the formula since it will be given in the formula booklet. (Statistics and probability)

Mutually exclusive (events)
In probability, things which cannot happen at the same time. (Statistics and probability)

Natural logarithm	A base e logarithm; the logarithm of a number using base e. (Number and algebra)
Natural logarithmic model (AHL)	Any relationship of the form $f(x) = a \ln(bx + c) + d$ (defined for $bx + c > 0$, where the independent variable x appears in the natural logarithm. Its **vertical asymptote** is the line $x = -\frac{c}{b}$. For $a > 0$ and $b > 0$, as well as for $a < 0$ and $b < 0$, the function is increasing, whereas for $a > 0$ and $b < 0$, as well as for $a < 0$ and $b > 0$, the function is decreasing. (Functions)
Natural number	Positive whole counting numbers. In IB, this includes 0; so natural numbers are 0, 1, 2, 3, 4, 5 …… (Number and algebra)
Navigate (navigation)	To plan or plot a course or journey using **bearings**. We use trigonometry in Mathematics to calculate these journeys. Remember that when you are measuring bearings, you measure from the north line, clockwise. (Geometry and trigonometry)
Nearest neighbour algorithm (AHL)	An **algorithm** used to find the **upper bound** of the **travelling salesman problem**. It selects a random **vertex** and finds the shortest path connecting it to any other unvisited vertex, then repeats at the next vertex until all vertices are visited. (Geometry and trigonometry)
Nearest neighbour graph	A **directed graph** where each **vertex** is connected by an **edge** to its nearest neighbour. (Geometry and trigonometry)
Nearest neighbour interpolation	A method of **interpolation** using the **nearest neighbour algorithm** in one or more dimensions. A **Voronoi diagram** is an example of nearest neighbour interpolation as it contains cells for each **site** where every point in the cell is nearer to that site than any other site. (Geometry and trigonometry)
Necessary condition	Something which has to satisfied for a statement to be true.
Negative (of a vector) (AHL)	A **vector** of the same **magnitude** but in the opposite direction. Also called opposite vector. (Geometry and trigonometry)
Negative correlation	A relationship between two variables where as one variable increases, the other decreases. (Statistics and probability) (see **positive correlation**)
Negative exponential	A power which is less than zero. So for a number $a^{-n} = \frac{1}{a^n}$ where $n > 0$, then a has a negative exponential. (Functions)
Network (as in graph theory) (AHL)	A collection of connected points (**vertices** connected by **edges**). (Geometry and trigonometry) (see **graph** (as in graph theory))
Newton's notation	Also called dot notation. Instead of dy/dx, we write a y with a dot over it, ie $\dot{y}$. (It is not commonly used.) (Calculus)
No correlation	When you have two sets of data that have no relationship to each other, you say that there is no correlation between them. For example, there is no correlation between data for size of shoes and marks out of 10 in a Mathematics test. (Statistics and probability)
Node	In **graph theory** a point on a graph, also called a vertex. For an example, see **graph** (as in graph theory). (Geometry and trigonometry)

N

Non-commutative (non-commutativity) (AHL) — The property stating that an **operation** on two or more objects depends on the order of the objects. For example, number addition and multiplication are both **commutative**:

$10 + 2 = 2 + 10 \qquad 10 \times 2 = 2 \times 10$ However, number subtraction and division are non-commutative:

$10 - 2 \neq 2 - 10 \qquad 10 \div 2 \neq 2 \div 10$ (Number and algebra)

Non-exclusive (non-exclusivity) — Events which can occur at the same time. (Statistics and probability)

(see **exclusive**, **inclusive**, **or**)

Non-included angle — An angle in a shape which is not between two given lines. (Geometry and trigonometry)

(see **included angle**)

Non-linear — When the relationship between two **variables** cannot be represented by a line (it may be a curve, etc). (Functions, Statistics and probability)

Non-linear regression (AHL) — A **regression** curve (curve of best fit) that takes the form of a function that is not a straight line (or **linear**). (Statistics and probability)

(see **linear regression**)

Non-singular (matrix) (AHL) — A matrix that has an **inverse (matrix)**. The **determinant** of a non-singular matrix is *not* equal to zero. (Number and algebra)

(see **singular matrix**, **inverse matrix**, **determinant**)

Non-trivial — In everyday language, this would mean not trivial, or significant. In Mathematics, **trivial** is often used to describe an example or solution involving zero, so non-zero solutions are seen as non-trivial.

Non-zero vector (AHL) — A **vector** whose **magnitude** is not zero. (Geometry and trigonometry)

Normal (AHL) — The line that is perpendicular to the tangent line at a point on a curve. Also called normal line. (Geometry and trigonometry, Calculus)

Normal CDF — A **GDC** function which helps to find the probability that a value will fall within specified boundaries in a normal distribution. (Statistics and probability)

Normal curve — The graph drawn of a normal distribution. (Statistics and probability)

Normal distribution (or normally distributed) — A symmetrical curve – often called 'bell-shaped' or bell curve. Blood pressure readings or heights of students have a normal distribution. (Statistics and probability)

Normal mode — A **GDC** mode in which numbers are not generally converted to scientific notation. (Number and algebra)

Normal PDF — A **GDC** function used for calculations for the **normal distribution**. Not generally used at SL. (Statistics and probability)

Normal probability (AHL) — The probability that a **normal random variable**, X, is less than or equal to a certain value, x, in a **normal distribution**, $P(X \leq x)$. For example, given that the heights of IB Diploma Mathematics students follow a normal distribution, you can find the probability that a student is at most a particular height. You must be able to find normal probabilities using your calculator. You should be able to calculate the probability that a random variable, X, lies in a certain range, $P(X \leq x)$ and if given a cumulative probability, find the boundary value, x. (Statistics and probability)

Normal random variable (AHL) — A **random variable** that follows the **normal distribution**. (Statistics and probability)

Normality (AHL)	When a **random variable** follows a **normal distribution** it is said to have normality. (Statistics and probability)
Normalize (normalization) (AHL)	To turn a **vector** into a **direction vector** (a vector with a magnitude of 1). To do this, divide each component of the vector by its **magnitude**. For example, to normalize the vector $v = 3i + 4j$, divide each component by the magnitude of v, $\|v\| = \sqrt{3^2 + 4^2} = 5$. The unit vector is then, $\frac{v}{\|v\|} = \frac{3}{5}i + \frac{4}{5}j$. (Geometry and trigonometry)
Notation	A system of writing and symbols used in Mathematics to represent a shortened form of a concept.
nth term	The way of describing a general term in a sequence of numbers. For example, the nth term of the arithmetic sequence that begins, 3,5,7,9,11 … is $3 + (n-1)2$. (Number and algebra)
nth term formula (*or* nth term rule)	A formula which allows you to find any number in a sequence of numbers. Remember that for any task as part of IB, the formulae for n**th terms** of arithmetic and geometric sequences are given on the formula sheet, so they are not something you have to remember. (Number and algebra)
Null hypothesis	The statement that is being tested in a **hypothesis test**. You then set out to determine whether there are statistical reasons for accepting or rejecting this hypothesis. (Statistics and probability)
Null vector (AHL)	(see **zero vector**) (Geometry and trigonometry)
Number line	A straight line representing positive and negative digits in their usual order, with zero in the middle. (Number and algebra)
Number system	A writing system to express numbers. Number systems change over time and in different cultures. Today, many countries use the **decimal number system**. (Number and algebra)
Numerator	The top part of a fraction. For example, in ½, 1 is the numerator. (Number and algebra) (see **denominator**)
Numerical	Expressed in or relating to numbers and symbols such as +, −, etc.
Numerical evaluation	The process of working something out using numbers. (Number and algebra)
Numerical integration	A method of **integration** that finds an **approximation** for the area under a curve using geometrical techniques. (Calculus) (see **trapezoidal rule**)
Numerical solution (of a differential equation) (AHL)	An approximate solution to a **differential equation** found by calculating a sequence of values that are close to the true solution. (Calculus) (see **Euler's method**)
Numerical value	An amount expressed in numbers. (Number and algebra)

O

Observed data	The resulting, recorded numbers from an experiment. In statistics, we can do calculations to test whether observed data matches expected results, such as by using the **chi-squared goodness of fit test**. (Statistics and probability)
Observed frequency	How often an event in an experiment occurs, such as the number of times a 6 is thrown when rolling a die 20 times. (Statistics and probability)
Obtuse angle	An angle which is larger than 90 degrees but less than 180 degrees. (Geometry and trigonometry)
Occurrence	Something that happens. (Statistics and probability)
Odd degree (of a vertex) (or odd vertex) (AHL)	A **vertex** that has an odd **degree**, ie it has an odd number of edges joining it. (Geometry and trigonometry)
Odd length (of a path) (AHL)	A **path** that has an odd number of edges. (Geometry and trigonometry
One to one function	Where each y value in the **range** corresponds to one and only one x value in the **domain**. For example, $y = x^2$ is not a one to one function because a y value such as 4 is produced by more than one x value, ie 2 and -2. (Functions)
One-dimensional (1D) (AHL)	Related to one dimension (**1D**) when we measure in one direction, usually length (a line). In calculus, used to describe an object that moves only in one direction.
One-tailed test	Used in statistical tests, such as a **t-test**, where you are interested in whether a calculated result is either greater than or less than a certain value. (Statistics and probability) (see **critical value**)
Open-minded	The IB learner profile suggests a number of qualities that students should be aiming for. One of these is to be 'open-minded' which can involve being open to and considering the views and ideas of others without prejudice. In Mathematics, this can mean working with others and listening to how they explain something even if it is different to the way you might normally do it.
Operation	A process that is applied to numbers. The main operations in Mathematics are addition, subtraction, multiplication and division. (Number and algebra)
Opposite	On the other side or across from it. Numerals which are the same but which have different signs, eg -4 and $+4$ are opposite numbers. In trigonometry, the side of the **right-angled triangle** which is across from the angle in which we are interested.
Opposite numbers	Two numbers whose sum is equal to 0, ie the additive identity element. In other words, if $a + b = 0 \Rightarrow b = -a$, or the **additive inverse** of a. One can say that b has the same **absolute value** as a, but the opposite sign. (Number and algebra)
Opposite vector (AHL)	(see **negative (of a vector)**) Geometry and trigonometry)
Optimization problem	A task or question which involves finding maximum and minimum values. (Calculus)
Optimize (optimization)	To find the maximum or minimum value. (Calculus)
Optimum position (on Voronoi diagram)	For a given set of at least three points, the position that minimizes the total distance from each point. If the set has exactly three points, this will be at the intersection of the **faces** of the **Voronoi diagram** corresponding to those points. (Geometry and trigonometry)

Or (in probability)	The probability of two events A or B happening is the probability of A happening, or B happening, or both. If the events A and B are mutually exclusive, the probability of A or B happening is the sum of the probability of A happening and the probability of B happening. (Statistics and probability) (see **exclusive**, **non-exclusive**, **inclusive**)
Order (of a graph) (AHL)	The number of **vertices** in a **graph**. (Geometry and trigonometry)
Order (of a matrix) (AHL)	(see **dimension (of a matrix)**) (Number and algebra)
Order of operations	This is often known by one of the acronyms BIDMAS, BODMAS or PEMDAS. These show you which of the mathematical rules or operations you should apply in turn in a calculation. For example, **brackets** (or parentheses) are worked out before **multiplying** (with BIDMAS). (Number and algebra)
Ordered pairs	Two numbers where order is important, eg (6, 8) is not the same as (8, 6). We write **coordinates** as ordered pairs to show where points appear on a grid. (Functions)
Ordered set	1. Two or more numbers arranged in a particular order $x_1, x_2, x_3 \ldots$. 2. Also used to describe the solutions in a system of equations. For example, the ordered set of numbers $-1, 3$ is the solution of the system $$\begin{cases} x - 2y = -8 \\ 3x + y = 0 \end{cases}$$ because the numbers $x = -1$ and $y = 3$ satisfy both equations. (Number and algebra)
Orientation	Which way round something is. For example, two triangles can be exactly the same as each other but not facing the same direction. The orientation of an object is often expressed in relation to the points of a compass. The earth's tilt (or orientation) is 23.5 degrees.
Origin	The starting point. Most often used to refer to the point (0, 0) on a grid. (Functions, Geometry and trigonometry)
Orthogonal (AHL)	At a right angle. (Geometry and trigonometry)
Out degree (AHL)	The number of **edges** for which a particular **vertex** is their initial (starting) vertex in a **directed graph**. (Geometry and trigonometry)
Outcome	What happens as a result of doing something, eg the outcome of throwing a coin can be heads or tails. (Statistics and probability)
Outlier	An amount which is outside the expected **range** of numbers for a data set. For example, a temperature of $38°C$ in a set of maximum daily temperatures for Iceland would be an outlier because it is outside the expected temperature range for Iceland. There is a formula from which you can calculate whether a number is an outlier. (It does not appear on your formula sheet.) (Statistics and probability)
Output (value)	An amount which is formed when a number is substituted into an expression. (Functions)
Overestimate	When the amount you roughly calculate is greater than the real value. (Number and algebra, Calculus)

P

Paired samples	When data is linked. For example, Asif's weight is 62kg and his height is 1.8m so 62 and 1.8 is a paired sample. (Statistics and probability) (see **bivariate data**)
Parabola	A curve which is both **symmetrical** and U-shaped. (Functions)
Parallel	Two straight lines are parallel if they stay the same distance apart and never meet, however long they go on for. (Functions)
Parallel forms (AHL)	Two forms of a test that both measure the same criteria so that each participant should score the same on both tests and the **mean** and **variances** of the two tests should be the same. The **correlation** between scores is used to measure the reliability of the test. For example, a group of students studying IB Diploma Mathematics sit an exam. An exam is written and split into two exams (Exam A and Exam B) that measure mastery of the same topics. Exam A is given to a sample of students on Monday and Exam B is given to those same students on Tuesday. The exams are parallel forms if the mean and variance are similar between the two exams. (Statistics and probability)
Parallelogram	A four-sided shape with two pairs of opposite, equal and **parallel** sides. (Geometry and trigonometry)
Parameter	A condition or restriction which helps to describe a system. (Number and algebra, Functions, Statistics and probability)
Parameter (of a population)	A characteristic (of a **population**). (Statistics and probability)
Parametric equations (AHL)	Equations that describe movement of a point along a curve where the x-, y- and z- coordinates are given in terms of functions of another variable. (Geometry and trigonometry)
Parametric form (of vector equation of a line) (AHL)	**Parametric equations** can be used to write equations for a straight line from a vector. Given a point travelling in a straight line, the vector equation of the line (also called the **position vector**) is $r = a + \lambda b$, where r is the position at time t, a is the starting position, λ is the time, and b is the direction vector. In parametric form, this becomes $x = x_0 + \lambda l$, $y = y_0 + \lambda m$, $z = z_0 + \lambda n$ (Geometry and trigonometry)
Parametric representation (AHL)	Presenting the **solution set** of a single linear equation using a **parameter**. More specifically, a linear equation of the form $ax + by = c$ is **satisfied** by an infinite number of ordered pairs (x, y), ie the points on the straight line representing this equation. If the equation is rewritten as $y = \frac{c-ax}{b}$, then using the parameter t to represent x, ie $x = t$, you can write the **solution set** of the equation as $\left(t, \frac{c-at}{b}\right)$, where the parameter t can take all real values. Note that you can also use $y = t$ to arrive at an equivalent parametric representation of the linear equation, eg $\left(\frac{c-bt}{a}, t\right)$. (Number and algebra)
Partial sum (of a geometric series) (AHL)	The sum of the first n terms of a geometric sequence, S_n, that is given by the direct formula $S_n = \sum_{k=1}^{n} u_k = \frac{u_1(1-r^n)}{1-r}, r \neq 1$. For the limit of the partial sum as $n \to \infty$ (for $|r| < 1$), see **infinite geometric series**. (Number and algebra)
Particular solution (AHL)	The **unique** solution to a **differential equation** when the equation is given additional constraints. Unlike the **general solution**, the particular solution does not have any **constants** in the equation. For example, the particular solution to the differential equation $\frac{dy}{dx} = 4x^3$ given the constraint that when $x = 1, y = 2$ is $y = x^4 + 1$ (Calculus)

Pascal's triangle	A triangular shape of numbers in which each number in the triangle is the sum of the two numbers above it. (Statistics and probability)
Path (AHL)	A route taken from one **vertex** to another (a **walk**) that does not revisit any vertices. It is a sequence of **edges** and **adjacent vertices**. (Geometry and trigonometry)
Path-find (on Voronoi diagram)	To find the shortest distance between two points. A **Voronoi diagram** is one way in which to path-find. (Geometry and trigonometry)
Pearson's product-moment correlation coefficient	A calculation which works out the strength of the linear relationship between two sets of **variables**. This can be easily calculated on a GDC. (Statistics and probability)
Percentage (*or* percent)	An amount out of 100 or parts per 100. (Number and algebra)
Percentage error	The error in an estimation, given as a **percentage** of the accurate solution. (Number and algebra)
Percentage point (of a *t*-distribution) (AHL)	For a given **random variable**, T, the percentage point is the value, t, for which $P(T \leq t)$ is equal to a given probability. If $P(T \leq t) = p\%$, then t is the pth percentage point. (Statistics and probability)
Percentile	A set of data divided into one hundred equal parts. (Statistics and probability)
Perfect correlation	There is an exact relationship between two variables. If this relationship is linear, the data when plotted forms a straight line and you would obtain a value of $+1$ or -1 for the **Pearson's product-moment correlation coefficient**. (Statistics and probability)
Perfect cube (*or* cube number)	A number that is the result of an integer raised to the power of 3 or multiplied by itself twice. For example, 64 is a perfect cube as $64 = 4 \times 4 \times 4 = 4^3$. (Number and algebra)
Perfect negative correlation	There is an exact inverse relationship between two variables, so that as one variable increases the other decreases. If this relationship is linear, the data when plotted forms a straight line with a negative gradient and you would obtain a value of -1 for the **Pearson's product-moment correlation coefficient**. (Statistics and probability)
Perfect positive correlation	There is an exact relationship between two variables, so that as one variable increases the other increases. If this relationship is linear, the data when plotted forms a straight line with a positive gradient and you would obtain a value of $+1$ for the **Pearson's product-moment correlation coefficient**. (Statistics and probability)
Perfect square (*or* square number)	When a number is the result of an integer raised to the power of 2 or multiplied by itself. For example, 36 is a perfect square as $36 = 6 \times 6 = 6^2$. (Number and algebra)
Perimeter	The distance around the outside of a shape, usually given in cm, m, etc. (Geometry and trigonometry)
Period	How far it is between the start of one wave of a graph until it repeats itself (or from any point to the next matching point). For example, for the graph of $y = \sin x$, it is 360 degrees. (Functions)
Periodic (function)	A graph, such as those for sine x and cosine x, which continues forever with a repeating pattern. The function returns to the same value at regular intervals (called a **period**). (Functions)

P

Periodic state (AHL) — A state in a **Markov chain** is said to be periodic if it is possible to return to itself and if the length of all paths returning to itself are multiples of a fixed number greater than one (the period). For example, states A and B in the Markov chain shown below are periodic because they can each be returned to in 2 steps. States C and D are not periodic as C cannot be returned to and D can only be returned to itself in 1 step. States C and D are known as **aperiodic states**. (Statistics and probability)

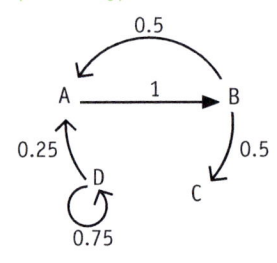

Perpendicular — At right angles to. When two lines are perpendicular the **angle** between them is 90 degrees. (Functions)

Perpendicular bisector — A line which is at right angles to another and cuts it exactly in half. You use a ruler and compasses if you wish to draw it accurately. (Geometry and trigonometry)

Petersen graph (AHL) — An undirected **cubic graph** that has 10 **vertices** and 15 **edges**. It is formed by connecting the vertices of a pentagon with the vertices of a star. (Geometry and trigonometry)

Phase portrait (AHL) — A graphical representation of the solutions of a **dynamical system** drawn as **parametric equations** following the path of each **particular solution**. A phase portrait can be used to show and predict the behaviour of the solution, look at the stability and to identify equilibrium and **saddle points**. (Calculus)

Phase portrait method (AHL) — Using a **phase portrait** to determine the behaviour of a **dynamic system**. (Calculus)

Phase shift (AHL) — Any **sine function** of the form $g(x) = A\sin(B(x - \emptyset)) + C$ can be obtained from the function $f(x) = A\sin(Bx) + C$ through a horizontal translation by $\emptyset$ units, where $\emptyset$ is referred to as the phase shift of g. Similarly, any **cosine function** of the form $h(x) = A\cos(B(x - \phi)) + C$, has a phase shift of $\emptyset$. (Number and algebra, Functions)

Phase shift angle (AHL) — (see **phase shift**) (Number and algebra)

Pi — The **irrational** number which is found by dividing the circumference of a circle by its diameter. It is usually approximated to 22/7 or 3.14. It is written using the symbol π in mathematical equations.

Pictogram — A simple graph where a picture represents an object. (Sometimes a picture can represent a number of objects (eg 5 or 10), so a key must be drawn to explain that.) (Statistics and probability)

Pie chart	A graph in the shape of a circle that is divided into sections from the centre to represent amounts and show the relative size of each amount. (Statistics and probability)
Piecewise linear model	A function that consists of two or more straight lines, each of them defined in mutually exclusive intervals, used to represent something mathematically. For example, $$v = \begin{cases} 2t, & 0 \leq t \leq 2 \\ 4, & t > 2 \end{cases}$$ could be used to model a car that accelerates (with constant acceleration) for 2 seconds and then maintains a steady speed. (Functions)
Piecewise model (AHL)	A **model** adopted to represent data that appear to be drawn from a **piecewise relationship**. When modelling data points that clearly appear to follow different trends in different parts of the domain they are drawn from, you can attempt to fit the different parts of the data set with different models, thus arriving at a piecewise model. If the data points extend continuously (they do not present gaps), you must pay attention when deciding the point where the piecewise model will change formula, so that the model is **continuous** at this point. (Functions)
Piecewise relationship (AHL)	A **relationship** that adopts different formulae in different parts of its **domain**, specifically $f(x) = \begin{cases} g(x), x \geq a \\ h(x). x < a \end{cases}$, where a is an arbitrary point in the relationship's domain $(a \in D_f)$. For example, $f(x) = \begin{cases} x - 2, & x \geq 2 \\ x^2 - 5x + 6, x < 2 \end{cases}$. In general, a piecewise relationship does not have to be **continuous** at the arbitrary point a where it changes formula. However, continuous piecewise relationships are particularly useful when modelling data points that follow different trends in different parts of the domain they are drawn from. In your IB Diploma exam, possible questions include providing a piecewise relationship that contains parameter(s) in its multiple formulae, where HL students are asked to find the value of the parameter(s) so that the relationship is continuous in its domain. These questions can be answered by forcing the different formulae of the relationship to be equal to each other at the point where the formula changes, eg $f(x) = \begin{cases} x - 2, & x \geq 2 \\ x^2 - kx + 6, x < 2 \end{cases}$, is continuous when $10 - 2k = 0$ (obtained when $x = 2$ is substituted into both formulae), ie $k = 5$. (Functions) (see **piecewise model**)
Pigeonhole principle (AHL)	The idea that if you are putting items into containers (ie pigeonholes) and you have more items than containers, at least one container will have more than one item. (Geometry and trigonometry)
Place value	Where an individual digit is positioned in a number, showing its value. (Number and algebra)
Planar (graph)	Flat, on a **plane**, or like a plane. A planar graph is flat and appears on a plane. (Geometry and trigonometry)
Planar form	Flat and fitting on a plane. (Geometry and trigonometry)
Planar graph (AHL)	A **graph** that can be drawn in a plane so that the **edges** do not cross each other. (Geometry and trigonometry)
Plane	A flat surface which goes on forever. (Geometry and trigonometry)
Plane figure	A **2D** shape. (Geometry and trigonometry)
Plot	Used when drawing graphs – to plot the point (3, 2) means to draw or put it in the correct place on the grid, graph or map. Often used as a **command term**.

P

Point
An exact location or position. **Coordinates** describe a point and its position. For example, $(0, 0)$ is a point called the **origin** and is where the x- and the y-axes of a graph cross. (Geometry and trigonometry)

Point estimate (AHL)
A formula that uses a sample statistic (eg **standard deviation**) to calculate an estimate of that statistic for the **population**. (Statistics and probability)

Point of inflexion (AHL)
A point where the curve changes from **concave up** to **concave down** (or from concave down to up). The **second derivative** of the function is zero, ie $f''(x) = 0$, at a point of inflexion and the sign of $f''(x)$ changes on either side of the point, ie $f''(x) < 0$ on one side of the point of inflexion and $f''(x) > 0$ on the other. (Calculus)

Point of intersection
The **coordinates** of where two lines meet (or intersect) each other. (Functions)

Point-gradient form (of a straight line equation)
When you know the steepness or **gradient** of a line, m, and a point through which it passes (x_1, y_1) the equation of the line is given by $y - y_1 = m(x - x_1)$. You can use this equation to find other points on the line. (Functions)

Poisson distribution (AHL)
A **probability distribution** that describes the number of events that will occur within a specified region (area or volume) or within a given time period. The Poisson distribution is scalable – eg if the mean number of events in one minute is m, then the mean number of events in three minutes is $3m$.

The Poisson distribution can be used to model situations that are counts of numbers of events in a particular region or time period; the probability of each event occurring must be the same for each region or time period (**uniform average rate**); and the number of times an event occurs in one region must be independent of the outcome in other regions. For example, the number of speeding fines issued per week by a certain speed camera can be modelled as a Poisson distribution. (Statistics and probability)

Poisson test (AHL)
A test that uses the **probabilities** from a **Poisson distribution** to determine how significantly data from an experiment differs from expected outcomes. To use the Poisson test, you need the hypothesised mean number of successes or events that occur in a specified region or time period, m, and the actual number of success that occur in the same region, x. Use the Poisson distribution to find the **probability** of the actual number of successes occurring by **chance** and compare it to the desired significance level. (Statistics and probability)

Polar form (of complex number) (AHL)
A complex number $z = x + iy$ (expressed in terms of its cartesian coordinates (x, y)) can be written in the form $z = r(\cos\theta + i\sin\theta)$. In other words, it can be expressed in terms of its polar coordinates (r, θ), ie its **modulus** and **argument**. (Number and algebra)

(see **modulus-argument form (of complex number)**)

Polygon
A **two-dimensional (2D) shape** with many sides, eg a pentagon has 5 sides, an octagon has 8. (Geometry and trigonometry)

Polyhedron (plural polyhedra)
Rather like a **three-dimensional (3D)** polygon – a **3D** shape with many faces. (Geometry and trigonometry)

Polynomial
An expression with several terms, usually the sum of different powers of the same variable, eg $5x^4 + 7x^3 - 4$. (Number and algebra)

Polynomial equation
An expression made up of several terms (numbers and variables) and which is equal to something, eg $5x^4 + 7x^3 - 4 = 0$. (Number and algebra)

Pooled two-sample t-test
A method for working out the variance of two **populations** when using a **t-test**. (Statistics and probability)

Population	The group which is being examined and has something in common. It may refer to a group of people, but can also describe data for other things that we can subject to statistical analysis. We may study a sample of the population as a way of finding out information about the group. (Statistics and probability) (see **sample**, **random**)
Population distribution	The pattern of how a **population** is spread out. (Statistics and probability)
Population growth	The increase in the number in a **population**. (Functions)
Population parameter	A characteristic of a **population**. (Statistics and probability)
Position vector (AHL)	A **vector** that gives the position of a point relative to a starting point. For example, point A (5, 3, 2) has position vector $$\overrightarrow{OA} = \boldsymbol{a} = 5\boldsymbol{i} + 3\boldsymbol{j} + 2\boldsymbol{k} = \begin{pmatrix} 5 \\ 3 \\ 2 \end{pmatrix}$$ A position vector can also describe the movement of a point. As the point moves, the position vector changes, $\boldsymbol{r} = \boldsymbol{a} + \lambda \boldsymbol{b}$, where $\boldsymbol{r}$ is the position at time λ, $\boldsymbol{a}$ is the starting position, λ is the time, and $\boldsymbol{b}$ is the direction vector. (Geometry and trigonometry)
Positive correlation	A relationship between two variables such that as one variable increases, so does the other. (Statistics and probability) (see **negative correlation**)
Positive exponential	A **power** which is more than zero. So for a number a^n where $n > 0$, then a has a positive exponential (Functions)
Power	Also called the **exponent** or the **index**. When we say 3 to the power of 4, we can write 3^4 or $3 \times 3 \times 3 \times 3 = 81$. (We can also say 81 is a power of 3.) (Number and algebra)
Power regression (AHL)	A **regression** curve (curve of best fit) that takes the form of a power function. (Statistics and probability)
Power relationship (AHL)	A **relationship** f between two variables x and y, such that $y = f(x) = ax^p$, where a, p are real number constants ($a, p \neq 0$, otherwise it becomes a trivial relationship). Strictly speaking, it is a polynomial relationship with a single term. The value of constant p determines the shape of the relationship's graph. A power relationship is not always a function, or in cases where it is a function, it is not always **one to one**. For example: $y = 3x^2$ $\qquad y = \frac{1}{2}x^{-2} = \frac{1}{2x^2}$ $y = -2x^{\frac{1}{3}} = -2\sqrt[3]{x}$ $\qquad y = -\frac{2}{3}x^\pi$ $y = -\frac{3}{2}x^{-\frac{3}{4}} = -\frac{3}{2\sqrt[4]{x^3}}$ (Functions)
Precise questioning	A way of asking questions so that understanding is maximized and **bias** avoided. (Statistics and probability)
Predator/prey model (AHL)	A system of nonlinear **coupled differential equations** describing the growth/decay of competing populations over time, eg a system of predator (hunters) and prey (hunted) animals. This is an application of **diagonalization** of the system matrix in terms of finding its **eigenvectors** and **eigenvalues**. (Number and algebra, Calculus)

P

Predict (prediction) — A typical **command term** in your IB studies, where you need to work out from information given what might be expected to happen next. Often used in sequences of numbers.

Prefix — A group of letters or word added in front of another word that changes its meaning eg **milli-** is the prefix in millimetre. (Number and algebra)

Prime factor — A number which divides exactly into another number without leaving a remainder and which is also a **prime number**. For example, the prime factors of 12 are 2 and 3. (Number and algebra)

Prime number — A number which has exactly two **factors**, one and itself. The number 1 is *not* prime. (Number and algebra)

Prim's algorithm (AHL) — An **algorithm** that finds a set of **edges** that form a **minimum spanning tree** connecting all the vertices. Prim's algorithm can also be used with matrices and tables. You should know how to apply it to a graph and to a matrix or table. (Geometry and trigonometry)

Principal (of a matrix) (AHL) — (see **diagonal (of a matrix)**) (Number and algebra)

Principal axis — A **horizontal** line halfway between the maximum and minimum points, eg for the sine graph $y = \sin x$, it is $y = 0$. (Functions)

Principled — The IB learner profile suggests a number of qualities that students should be aiming for. One of these is to be 'principled' which involves being honest, whilst considering the possible consequences of your actions and making sure that all you do is ethical. In Mathematics, this could mean that you work through a problem, even though you know it would be easy to look at a mark scheme or the answer pages. It's important to use these in the right way, not simply to copy from.

Prism — A solid **three-dimensional shape** which has the same shape going all the way through it and takes its name from the shape of its base. For example, a triangular prism has a triangle as its base. If it were a loaf of bread, wherever you cut it, you would get the same shape. (Geometry and trigonometry)

Probability — The **chance** or likelihood of something happening. Probabilities are always numbers written between 0 and 1. So, when throwing a coin, the probability of it landing on heads is 0.5. (Statistics and probability)

Probability distribution — The set of **values** which are possible as a result of an experiment. (Statistics and probability)

Probability vector (AHL) — A vector whose entries are probabilities that sum to 1. Probability vectors are used to calculate probabilities in **Markov chains**. (Statistics and probability)

Product — The product of two numbers is what you get when you multiply them together. The product of 4 and 5 is 20 as $4 \times 5 = 20$. (Number and algebra)

Product rule (AHL) — The method of finding the derivative of an expression that can be written as a **product** (multiplication). For $y = f(x)g(x)$, then $\frac{dy}{dx} = f'(x) \times g(x) + f(x) \times g'(x)$. The product rule is given in the formula booklet so you do not need to memorize it. (Calculus)

Profit — The amount gained when something is sold after all expenses have been taken away. For example, I buy a jacket for £59 and sell it for £65 so I make £6 profit. (Number and algebra)

Profit function, $P(x)$	Provides the **profit** made from selling a product/service as a function of the number of items sold, x. The profit function is calculated as the difference between the **income function** and the **cost function**, ie $P(x) = I(x) - C(x)$. (Functions)
Progression	A **sequence** of numbers which are formed by adding the same number (**common difference**) or multiplying by the same number (**common ratio**) each time. (Number and algebra)
Projectile (AHL)	An object thrown through the air. (Geometry and trigonometry)
Projectile motion (AHL)	The movement of an object thrown through the air. (Geometry and trigonometry)
Proportion	Where two ratios (or fractions) are equal. (Number and algebra)
Proportional	Two **variables** x and y are said to be proportional if there is a constant m such that $y = mx$. (Number and algebra)
Prove (proof)	A typical **command term** used in your IB Diploma studies which means that you need to show that something is true.
p-value	A figure used in hypothesis tests such as the **chi-squared goodness of fit test** with which you can compare with the **significance level** to see whether a **hypothesis** should be accepted or rejected. (Statistics and probability)
Pyramid	A **three-dimensional shape** which usually has a base of a square or triangle, from which all edges taper to meet at a point. (Geometry and trigonometry)
Pythagoras' theorem	A method which enables you to work out the length of any side of a right-angled triangle, given the lengths of the other two sides. Written as $a^2 + b^2 = c^2$. (Geometry and trigonometry)
Pythagorean identity (AHL)	The trigonometric identity is $\sin^2\theta + \cos^2\theta \equiv 1$ This identity can be derived using **Pythagoras' theorem** so is sometimes called the Pythagorean identity. The associated identities $1 + \tan^2\theta \equiv \sec^2\theta$ $1 + \cot^2\theta \equiv \csc^2\theta$ are also sometimes called Pythagorean identities. You will be given these identities in the formula book. (Geometry and trigonometry)

Q

Quadrant
1. One of the four equal parts of a circle when it is divided into four by two perpendicular diameters. (Geometry and trigonometry)
2. One of the four regions defined on the Cartesian plane by the coordinate axes. Points in the first quadrant have both positive x and y coordinates. The remaining quadrants are numbered counter-clockwise starting from the first. So, for example, points in the second quadrant have negative x and positive y coordinates. (Geometry and trigonometry)

Quadratic equation (AHL) An equation of the form $ax^2 + bx + c = 0$, where a, b, c are real number constants ($a \neq 0$, otherwise it becomes a linear equation).

The equation has two **solutions** (or **roots**) of the form $x = \frac{-b \pm \sqrt{b^2 - 4ac}}{2a}$. The nature of these solutions is determined by the sign of discriminant $\Delta = b^2 - 4ac$.

If $\Delta > 0$, there are two **distinct** real roots, representing the **x-intercepts** of the parabolic graph of the corresponding quadratic function $f(x) = ax^2 + bx + c$.

If $\Delta = 0$, the real root $x = \frac{-b}{2a}$ is a repeated (double root), representing the **vertex** of the parabolic graph that lies on the **x-axis**.

If $\Delta < 0$, the parabolic graph does not cross the x-axis. This means that the equation has no real roots. Instead, the roots of the equation are now two **conjugate complex numbers** of the form $x = \frac{-b \pm i\sqrt{4ac - b^2}}{2a}$. These numbers represent points on the **Argand diagram** that lie on the axis of symmetry of the corresponding parabola $\left(x = \frac{-b}{2a}\right)$, being $\frac{i\sqrt{4ac - b^2}}{2a}$ away and on either side of the x-axis. (Functions)

Quadratic expression (AHL) Any **algebraic expression** that has the form, or can be simplified to the form, $ax^2 + bx + c$, where a, b, c are real number constants ($a \neq 0$, otherwise it becomes a linear expression). Specifically, any expression that takes the form of a second-degree polynomial. For example, $-x^2 + 6x - 17$. (Functions)

Quadratic function A polynomial relationship where the largest term is a multiple of x^2, for example, $f(x) = 5x^2 + 7$. (Functions)

Quadratic inequality (AHL) An **inequality** involving a second order **polynomial**. Such an inequality is formed when the = sign is replaced by an inequality sign $\leq, <, >,$ or $\geq$. For example:

$ax^2 + bx + c \geq 0, a \neq 0$

A quadratic inequality can be solved either algebraically or graphically. In general, one needs to identify the intervals at which the **quadratic expression** takes positive or negative values. (Number and algebra, Functions)

Quadratic model A way of representing a real-life problem mathematically, using quadratic (x^2) values. (Functions)

Quadratic regression (AHL) A **regression** curve (curve of best fit) that takes the form of a **quadratic function**. (Statistics and probability)

Quadrilateral A four-sided two-dimensional shape. (Geometry and trigonometry)

Qualitative (data) Information that describes something and which can be arranged in categories but that isn't associated with numbers, eg the colour of the sea. (Statistics and probability)

(see **quantitative (data)**)

Quantify (quantification) To measure how many there are of something.

Quantitative (data)	Information which is numerical or about quantities and which can be measured, eg how many siblings people have. (Statistics and probability) (see **qualitative (data)**)
Quantity	An amount that can be measured, eg the quantity of flour needed is 400 grams.
Quarterly	Something that happens four times a year, or every quarter. For example, we might talk about **interest** being calculated quarterly. (Number and algebra)
Quartile	Dividing something into 4. You can have the lower quartile at 25% and the upper quartile at 75%. (Statistics and probability)
Questionnaire (AHL)	A list of questions used to collect data from respondents in a **survey**. Responses may be text, numerical, or multiple choice, and questions may be closed or open-ended. (Statistics and probability)
Quotient	The result when you divide a number by another number. (Number and algebra)
Quotient rule (AHL)	The method of finding the **derivative** of an expression that can be written as a **quotient** (division or fraction). For $y = \frac{f(x)}{g(x)}$, then $\frac{dy}{dx} = \frac{f'(x) \times g(x) - f(x) \times g'(x)}{(g(x))^2}$ The quotient rule is given in the formula booklet so you do not need to memorize it. (Calculus)

R

Radian (AHL)	A unit for measuring angles as an alternative to **degrees**. On a unit circle (a circle with radius 1), an arc of length 1 is delimited between two **radii** forming an angle of 1 radian. Hence, a full angle of 360° corresponds to a radian angle of 2π. Therefore, the conversion factor between degrees and radians is $\pi/180°$. The formulae for calculating the length ($l = \theta r$) and the area ($A = \frac{\theta}{2}r^2$) of a sector in a circle involve the angle θ in radians. In IBDP HL Mathematics, unless otherwise stated, you should assume angles are measured in radians. (Functions, Geometry and trigonometry)	

Radian measure (AHL) — The measure of an angle in **radians**. Angles in radians can be expressed as decimal numbers (eg 1.2, 2.5) or as multiples of π (eg π/2, π/3, 3π/2, 5π/6, etc). (Functions, Geometry and trigonometry)

Radian mode — A **GDC** function when working with angles. Not generally used for most topics at SL as **degree mode** is more usual (check on your mode before you start **trigonometry** as you will get the wrong answer if it is set to radian mode.)

(see **degree mode**)

Radical — A number or expression containing the symbol √ such as a square root or a cube root. (Number and algebra)

(see **surd**)

Radius (*plural* radii) — A straight line which goes from the centre of a circle to the **circumference**, or from the centre of a **sphere** to any point on its surface. (Geometry and trigonometry)

(see **diameter**)

Random experiment — A trial where one does not know the **outcome**. (Statistics and probability)

Random sample (random sampling) — A sample chosen entirely by **chance** (or at random), so that each member of the population has an equal opportunity of being chosen. (Statistics and probability)

Random variable — The result from a **random sample**. (Statistics and probability)

Range — The difference between the largest number and lowest number in a **data set**. To find the range, subtract the smallest number from the largest. (Statistics and probability)

Range (of a function) — The values taken by the dependent variable. For a function $y = f(x)$, the y values. (Functions)

Rank — To put a list of numbers in numerical order or in a sequence and give labels eg 'first', 'second', etc. (Statistics and probability)

Rate of change — How one **variable** changes with respect to another. (Calculus)

Ratio — A numerical relationship between two quantities. For example, if Harry is 26 and Ameera is 12, the ratio of their ages is 26:12 which can be simplified to 13:6. (Number and algebra)

Rational — A number which can be expressed as a **decimal** or **fraction**. The most common examples of numbers which are *not* rational are π, $\sqrt{2}$ and $\sqrt{3}$. (Number and algebra)

(see **irrational**)

Rational exponent — A **power** which can be expressed as a **decimal** or **fraction**. (Number and algebra)

Rationalize the denominator (AHL)	Remove all **radicals** from the denominator, by following specific steps, in order to write the fraction in a simplified form. For example: $$\frac{2}{1+\sqrt{5}} = \frac{2(1-\sqrt{5})}{(1+\sqrt{5})(1-\sqrt{5})} = \frac{2(1-\sqrt{5})}{1^2 - (\sqrt{5})^2} = \frac{2(1-\sqrt{5})}{-4} = \frac{\sqrt{5}-1}{2}$$ (Number and algebra)
Real number	All numbers that you know. These are every **rational** and every **irrational** number. The opposite of real numbers are imaginary numbers eg $\sqrt{-1}$, although these are not studied at SL. (Number and algebra)
Real part (of complex number) (AHL)	The part of a complex number that is *not* multiplied by the **imaginary unit**. For example, in a complex number $z = x + iy$, $x = \text{Re}(z)$ is the real part of z. (Number and algebra) (see **Cartesian form (of complex number)**, **imaginary part (of complex number)**)
Real solution (AHL)	A **real number** that **satisfies** an **algebraic equation**. For example, the equation $$(x^2 + 1)(x - 1) = 0$$ has three **distinct** solutions: $x = i$, $x = -i$ and $x = 1$. The first two are the **imaginary** solutions and the third is the real solution. (Number and algebra)
Real value	(see **real number**) (Number and algebra)
Rearrange (formula *or* expression)	To change the order of terms in a formula or expression. In a formula, we may want to change its subject. (Number and algebra)
Reasonable (reasonableness)	Something that is within the bounds of possibility. For example, the age of a man being 38 is reasonable. If his age is given as 238, that would not be reasonable.
Reciprocal	1 divided by a number gives the reciprocal of that number eg the reciprocal of 2/3 is 3/2 or 1.5.
Recurrent state (AHL)	A state in a Markov chain where the probability of returning to that state is 1. For example, state C is a recurrent state in the Markov chain shown below because wherever the process starts, it is certain to return to C. State B is a **transient state** because there is a non-zero probability that the system will not return to state B. (Statistics and probability)
Recurring decimal	A number with a **decimal** point which repeats for ever. This is often shown as '...'. For example, 1/3 is 0.333333333333 ... or 0.3 recurring. (Number and algebra)
Recursive formula	Where each term depends on the term before. (Number and algebra)

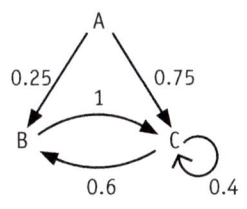

R

Reflection — As something would be seen in a mirror. We can have a line of reflection which would be a mirror line. (Functions, Geometry and trigonometry)

Reflective — One of the qualities from the IB learner profile to describe someone who thinks deeply about something. In your own study of Mathematics, this can mean you thinking about your own strengths and weaknesses and how you can use them and improve upon them.

Reflex angle — An angle which is larger than 180 degrees but less than 360 degrees. (Geometry and trigonometry)

Regression — The attempt to find a relationship between two variables. This can be done by putting a line of best fit on a scatter diagram or by calculation. (Statistics and probability)

Regression line (of y on x) — Used in statistics for showing the **gradient** and **intercept** of the line which best fits the data. The line describes how the dependent variable y changes as the independent variable x changes. (Statistics and probability)

Regular graph (AHL) — A graph in **graph theory** where each **vertex** has the same degree, ie each vertex has the same number of **edges**. (Geometry and trigonometry)

Regular Markov chain (AHL) — When the **transition matrix** of a **Markov chain** is regular, ie when there is a power of the transition matrix, T, which only has positive non-zero entries in the **matrix**. A regular Markov chain has a unique stationary matrix and each successive state matrix approaches the stationary matrix. (Statistics and probability)

Reject (rejection) — In a **hypothesis test**, you reject the **null hypothesis** when the calculated value is in the **critical region**. For example, in a **chi-squared test for independence**, when the calculated value is greater than the **critical value**, you discard (or reject) the null hypothesis. (Statistics and probability)

Related rates of change (AHL) — The rates of change of two or more variables that are related by time. For example, you can look at the position of a rocket after launch in the x- and y-direction and then find the **velocity** (the rate of change) in each direction. (Calculus)

Relationship — A link between two aspects of Mathematics. For example, x and y may have the relationship that $y = 2x + 3$.

Relative extremum (*plural* **relative extrema**) — See **local maximum** and **local minimum**. (Calculus)

Relative frequency — The number of times a certain outcome appears in an experiment as a fraction of how many times the experiment has been done. (Statistics and probability)

Relative maximum point — An alternative expression for **local maximum**. (Calculus)

Relative minimum point — An alternative expression for **local minimum**. (Calculus)

Relative position (AHL) — The position of A in relation to the position B. For example, suppose person A and B are standing 3 metres apart. Person A is on the left and person B is on the right. The relative position of person A is 3 metres to the left of person B. (Geometry and trigonometry)

Relevant (data) (AHL) — Data that is useful and applicable to the study. (Statistics and probability)

Reliability test (AHL) — Measures the consistency of the data. Data that is reliable will be accurate, reproducible, and consistent. Reliability can be tested through the **test-retest** method or through **parallel forms**. (Statistics and probability)

Reliable (reliability)	To describe whether data or the source it is taken from can be trusted. At AHL, you would be expected to know the difference between reliability and **validity**. (Statistics and probability)
Repayment	An amount of money paid back when money is borrowed or to repay a **debt**. It can be repaid per month, per year, etc. (Number and algebra)
Represent (representation)	To symbolize, exemplify, or take the place of something, eg in a pictogram, 1 picture of a cat may be used to show (or represent) 4 cats.
Representative (sample)	A sample taken which is in no way **biased** and which represents the whole **population** fairly. (Statistics and probability)
Rescale (rescaling) (AHL)	To convert a vector to a **direction vector** (also called a unit vector). To rescale a vector, v to a direction vector, divide the vector by its **magnitude**: $\frac{v}{|v|}$. (Geometry and trigonometry)
Residuals (AHL)	The difference between the data values (the points on a scatter diagram) of the dependent variable and the line or curve of best fit. For example, the line of best fit (**least squares regression curve**) of the points (2,2), (4,4), (6,6), (8,6), (10,8) is $f(x) = 0.7x + 1$. The residual of (2,2) is $2 - f(2) = 2 - 2.4 = -0.4$. (Statistics and probability)
Restriction (on growth) (AHL)	(see **logistic function**) (Functions)
Resultant (AHL)	The **vector** that is the sum of two other vectors. For example, the resultant of the vectors $3i + 4j - 2k$ and $4j - i$ is $3i + 4j - 2k + 4j - i = 2i + 8j - 3k$. (Geometry and trigonometry)
Revenue	Money raised, particularly as a result of an investment or from sales. (Number and algebra)
Rhombus	A four-sided shape whose sides are all of equal length and which has two pairs of equal opposite angles. (Geometry and trigonometry)
Right angle	An angle which is exactly 90 degrees. (Geometry and trigonometry)
Right cone	A cone whose **vertex** is immediately above the centre of its base. (Geometry and trigonometry)
Right-angled triangle	A three-sided **2D** shape where one angle is exactly 90 degrees. (Geometry and trigonometry)
Right-hand screw rule (AHL)	A rule used to determine the direction of the **vector product**. To find the direction of $a \times b$, line up the index finger of your right hand with vector a and your middle finger with vector b. Whichever way your thumb faces is the direction of the vector product, $a \times b$. (Geometry and trigonometry)

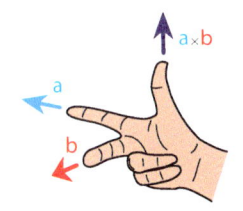

R

Right-pyramid — A pyramid whose highest **vertex** is immediately above the centre of its base. (Geometry and trigonometry)

Risk — Another word for **chance** or **probability** (usually of something unpleasant happening), eg the risk of losing an **investment**. (Statistics and probability)

Risk-taker — One of the qualities from the IB learner profile to describe how you might look at situations with which you are not familiar. In your study of Mathematics, this can mean having a try at a question which looks difficult and being prepared to get it wrong. (Sometimes this can help you learn.)

Root — The root of a number is something which when multiplied by itself one or more times gives that number, eg $\sqrt{5}$ is a root of 5 because $\sqrt{5} \times \sqrt{5} = 5$. (Number and algebra)

Root (of an equation) — A value that satisfies an equation. (Number and algebra, Functions)

(see **zero (of a function)**)

Rotation — A circular movement where the central point stays fixed. Rotation is usually described by the amount of degrees and direction of the turn and the **coordinates** of where the rotation starts from, eg a rotation of 90 degrees clockwise, centre (2,3). (Geometry and trigonometry)

(see **transformation**)

Rounded — Where a number or an answer is subject to **rounding** up or rounding down and given to a certain number of decimal places or significant figures. (Number and algebra)

Rounding — Making a number simpler, perhaps to a certain number of decimal places or significant figures, by either rounding up or rounding down. The number is less accurate, but easier to work with. (Number and algebra)

Rounding error — The difference between an exact answer and an answer which has been given to a number of decimal places or significant figures. (Number and algebra)

Route — The way to get from one place to another. (Geometry and trigonometry)

Row (of a matrix) (AHL) — A horizontal line of **elements** in a **matrix**. For example:

$$\begin{pmatrix} 1 & 2 & 3 \\ 3 & 0 & 1 \\ 0 & 1 & 2 \end{pmatrix} \quad \begin{pmatrix} 0 \\ 1 \\ 2 \end{pmatrix} \quad (0 \quad 1 \quad -1)$$

In some cases, a row may consist of only one element. In general, a matrix of the form

$$A = \begin{pmatrix} a_{11} & a_{12} & \cdots & a_{1n} \\ a_{21} & a_{22} & \cdots & a_{2n} \\ \vdots & \vdots & \ddots & \vdots \\ a_{m1} & a_{m2} & \cdots & a_{mn} \end{pmatrix} \begin{array}{l} \text{row 1} \\ \text{row 2} \\ \vdots \\ \text{row } m \end{array}$$

has rows with n **elements**. (Number and algebra)

(see **column (of a matrix)**)

Row vector (AHL) — A **matrix** consisting of one **row** and n **columns**.

$(a_1 \quad a_2 \quad \cdots \quad a_n)$

For example, a 1×3 row vector, such as $(2 \quad -1 \quad 4)$ can represent the position of a fly in a **3D** space. (Number and algebra)

(see **column vector**)

Saddle point (AHL)	An unstable point in the **phase portrait** of a dynamic system where solutions are unstable in one direction and stable in another. The origin is a saddle point if the **eigenvalues** are real with opposite signs. (Calculus)
Sample	A part or section of a **population** selected to **represent** the whole population. (Statistics and probability)
Sample space	A list of all the values in the **population** or all possible outcomes of an experiment, often shown by the symbol S, U or Ω. (Statistics and probability)
Sample space diagram	A table with all the possibilities for pairs of outcomes in an experiment. (Statistics and probability)
Sample statistic	A piece of data obtained from a **sample** of the **population**, such as the mean. (Statistics and probability)
Sampling	The process of taking some pieces of data from some members (a sample) of a **population**. (Statistics and probability)
Sampling techniques	Methods of choosing a sample from a **population** to represent the whole population. (Statistics and probability)
	(see **random sampling**)
Satisfy (satisfied)	Where a value (or set of values) makes an equation true. For example, $x = 1$ satisfies the equation $5x + 2 = 3x + 4$ (as $7 = 7$), or the equation $5x^2 - 3x - 2 = 0$ (as $0 = 0$). The **ordered pair** (1, 2) satisfies the equation $5x - 2y = 1$ (as $1 = 1$). (Number and algebra)
Scalar (AHL)	A quantity that can be described by a real number. A scalar quantity has size (or **magnitude**), but no direction. Examples of scalar quantities are mass, time, length, voltage, temperature, speed (magnitude of a velocity vector) etc. (Number and algebra, Geometry and trigonometry)
Scalar multiplication (AHL)	Multiplying a **vector** by a **scalar**. For example, given vector $\boldsymbol{u}$, the vector $3\boldsymbol{u}$ is said to be a scalar multiple of $\boldsymbol{u}$. (Geometry and trigonometry)
Scalar product (AHL)	An operation on two **vectors** that gives a **scalar** (a number, not a vector) as an answer. It is denoted with the symbol · and is said 'a dot b'. It is also called the dot product (or the inner product).
	$\boldsymbol{a} \cdot \boldsymbol{b} = \|\boldsymbol{a}\|\|\boldsymbol{b}\| \cos \theta$ where θ is the angle between $\boldsymbol{a}$ and $\boldsymbol{b}$. The scalar product of two perpendicular vectors is zero. (Geometry and trigonometry)
Scale factor	The number by which something is multiplied. For example, if a length of 8 cm is enlarged by a scale factor of 2.5, it becomes 20 cm, as $8 \times 2.5 = 20$. (Functions, Geometry and trigonometry)
	(see **enlargement**, **ratio**, **stretch**)

Scalene	Used to describe a triangle which has no equal sides or equal angles. (Geometry and trigonometry)
Scaling (AHL)	Where it is necessary to analyse **bivariate data** which has values that extend over several orders of magnitude (ie several powers of 10), we can use **logarithms** to scale these data. If both the independent and the dependent variable need scaling, then we can plot the logarithms of both quantities on a **log-log graph**. If only one of the two variables needs scaling, then we can plot the resulting logarithmic quantity against the other variable on a **semi-log graph**. If the data points on a **log-log** or a **semi-log plot** appear to be varying linearly, then we can obtain their **best-fit straight-line** model. If the latter indicates a strong linear relationship then conclusions can be drawn on the nature of the relationship of the original data, ie that before scaling. (Functions)
	(see **linearization**)
Scatter diagram	A graph where points are plotted to represent two variables that might be linked in some way. For example, you might plot the height and weight of 100 people, where each person is represented by a point on the graph, to see if there is a relationship between a person's height and their weight. (Statistics and probability)
Schlegel diagram (AHL)	A projection of a 3D shape into a 2D **planar graph**. (Geometry and trigonometry)
Sci	An abbreviation for **scientific mode** on a **GDC**.
Scientific mode	A mode on a calculator which makes numbers appear in **standard form** (as a number between 1 and 10 times 10 to a power) (or **scientific notation**). (Number and algebra)
Scientific notation	A way of expressing a number as a number between 1 and 10, multiplied by 10 to the power of something. Often used for very large or very small numbers. For example, 53000 in scientific notation would be 5.3×10^4. Also called **standard form**. (Number and algebra)
Scroll	To move down or up the screen on a calculator.
Second derivative (AHL)	The **derivative** of the derivative of a function. It is written as either $f''(x)$ or $\dfrac{d^2y}{dx^2}$ if $y = f(x)$. (Calculus)
Second derivative test (AHL)	A method of determining whether a point is a **maximum**, **minimum**, or **point of inflection**.
	Given a function $f(x)$ and a point a where $f'(a) = 0$ (the first derivative is zero), then if $f''(a) > 0$, then $f(a)$ is a minimum value of $f(x)$; if $f''(a) < 0$, then $f(a)$ is a maximum value of $f(x)$; if $f''(a) = 0$ and the sign of $f''(x)$ changes on either side of the point, then $f(a)$ is a point of inflexion. (Calculus)
Second order differential equation (AHL)	A **differential equation** that uses the **second derivative** of a function. (It can also use the **first derivative**, but must not use any derivative higher than the second.) For example, $\dfrac{d^2y}{dx^2} + 3\dfrac{dy}{dx} = 0$ is a second order differential equation because it contains $\dfrac{d^2y}{dx^2}$ as the highest derivative. (Calculus)
Sector	A part of a circle bounded by an **arc** and two **radii**. (Geometry and trigonometry)
	(see **segment**)

Seed (on Voronoi diagram)	(see **site (on Voronoi diagram)**) (Geometry and trigonometry)
Segment	A part of a circle bounded by an **arc** and a **chord**. (Geometry and trigonometry) (see **sector**)
Selling rate	The amount for which something is sold. Often used to refer to the rate at which a bank sells foreign currency. (Number and algebra)
Semi-Hamiltonian graph (AHL)	A graph that has a **path** that goes through every **vertex** exactly once but starts and ends at different vertices. (Geometry and trigonometry)
Semi-log graph (AHL)	In the analysis of **bivariate data**, in the case that the values of one variable extend over several orders of magnitude, one can use **scaling** to obtain the logarithm of the corresponding quantity, which can then be plotted on what is known as a semi-log graph (as the other variable is plotted on its original, non-logarithmic scale). (Functions)
Separable differential equation (AHL)	A differential equation that can be separated into two functions, each involving one **variable**, ie $\frac{dy}{dx} = f(x) \, g(y)$ (Calculus)
Separation of variables (AHL)	The method for solving **separable differential equations**. Step 1: Separate the x and y terms to opposite sides of the equation using algebra. Step 2: Treat $\frac{dy}{dx}$ as a fraction and separate it so that dy is on the side of the equation with the y terms and dx is on the side with the x terms. Step 3: **Integrate** both sides of the equation. Step 4: If the question asks for the equation in a particular form, rearrange the equation (eg as a function of y in terms of x). (Calculus)
Sequence	A set of numbers where each one is multiplied by or added to a certain number to find the next one. (Number and algebra) (see **arithmetic sequence**, **geometric sequence**)
Series	The sum of the numbers in a **sequence**. (Number and algebra)
Set	A grouping of objects or numbers which have something in common. (Number and algebra, Functions)
Show	Often used as a **command term** which means you are asked to prove something through clear steps in your workings.
Show that	A **command term** often used in your IB studies where you are asked to arrive at a result but without necessarily having to prove it. You would not normally need to use your calculator.
SI (Système International) unit	The modern metric system of measuring used throughout the world, where the units are m, km, kg, seconds, etc. (Number and algebra)
Sigma notation	A form of notation which signifies the sum of a group or sequence of numbers. It is written as the Greek letter Σ. (Number and algebra)
Significance level	In a hypothesis test, the number with which a value is compared to see whether the **null hypothesis** should be accepted or rejected. Expressed as a percentage or as a decimal. (Statistics and probability) (see **hypothesis test**)
Significant digit (or significant figure)	Generally, the digit which gives the most meaning to a number. In a decimal fraction, the first non-zero figure is the first significant figure. Used in **rounding**. (Number and algebra)

Simple event		Where one thing or experiment happens at a time producing a single outcome, eg tossing a coin. (Statistics and probability)
Simple graph (AHL)		An unweighted, **undirected graph** that has no loops and does not have more than one **edge** between any two **vertices**. (Geometry and trigonometry)
Simple interest		A method of calculating the amount of interest added when money is invested or borrowed, expressed as a percentage. Calculated only on the original amount, unlike **compound interest**. (Number and algebra)
Simple random sampling		A method of conducting an experiment where the sample is entirely chosen by **chance** (or at random). Useful where each result has an equal chance of occurring. (Statistics and probability)
Simplify (simplification)		To put something in the most straightforward way possible. For example, in algebra we simplify $5x + 3y - 3x + 7y$ to give $2x + 10y$. Or we can simplify the fraction $\frac{4}{8}$ to give ½. (Number and algebra)
Simulation		A mathematical model created to estimate or predict events or solve real-life problems.
Simultaneous		Happening at the same time. (Statistics and probability)
Simultaneous equations		A pair of equations involving two variables (at SL) which need to be solved at the same time, so the solutions are true for both equations. (Number and algebra)
Sine (sin) function		A trigonometric function. The sine of an acute angle can be between 0 and 1 and a graph can be drawn to represent it. (Functions)
		(see **cosine (cos) function, tangent (tan) function**)
Sine (sin) ratio		The opposite side divided by the length of the hypotenuse in a right-angled triangle. (Geometry and trigonometry)
		(see **cosine (cos) ratio, tangent (tan) ratio**)
Sine regression (AHL)		A **regression** curve (curve of best fit) that takes the form of a **sine function**. (Statistics and probability)
Sine rule		A method used to find an angle or side in a non-right-angled triangle. You must already know the size of one side and the angle opposite it. (Geometry and trigonometry)
		(see **sine rule**)
Singular (matrix) (AHL)		A matrix that does not have an **inverse (matrix)**. The **determinant** of a singular matrix is equal to zero. For example: $\begin{pmatrix} -1 & 1 \\ 2 & -2 \end{pmatrix}$ is singular, because $\begin{vmatrix} -1 & 1 \\ 2 & -2 \end{vmatrix} = 0$ (Number and algebra)
		(see **non-singular (matrix)**)
Sinusoidal		In the form of a sine curve, eg $\sin x$ or $\cos x$. (Number and algebra, Functions)
Sinusoidal function		A function which takes the form of a sine curve. (Number and algebra, Functions)
Sinusoidal model		A way of representing a real-life situation in mathematical terms, when the model involves a curve in the shape of a sine curve. (Functions)
Site (on Voronoi diagram)		A point that is placed on a plane from which a **Voronoi diagram** is made. Also known as a seed (on a Voronoi diagram). (Geometry and trigonometry)
Size (of a graph) (AHL)		The number of **edges** in a graph. (Geometry and trigonometry)

Sketch	To draw a diagram but not as roughly as you might do in an art class. In Mathematics, when you are asked to sketch, you must use a ruler for lines, but lines and angles do not have to be drawn exactly to scale. Often used as a **command term**.
	(see **draw**)
Skew lines (AHL)	Lines in three-dimensions that are not in the same plane. They do not intersect but also are not parallel. (Geometry and trigonometry)
Slope field (AHL)	A graphical representation of the solutions of a **differential equation**. It is a plot of the tangent lines (drawn as short line segments) at particular points of a differential equation. (Also called direction field or vector field.) (Calculus)
Slope function	The function obtained when you **differentiate** to find the steepness of a line or of a curve at a certain x value. (Calculus)
Solid	A **three-dimensional** shape, eg **sphere**, **cone**. (Geometry and trigonometry)
Solution	The answer to a question. (Number and algebra)
Solution curve (AHL)	The graph showing solutions to a **differential equation**. An approximate solution curve is tangent to the line segments of the **slope field**. The curve is determined by given initial conditions. (Calculus)
Solution set (AHL)	The set of all values **satisfying** an equation or expression. For example, in a single linear equation of the form $ax + by = c$, the solution set consists of the infinite number of ordered pairs $\left(x, \frac{c-ax}{b}\right)$, ie the points on the straight line representing this equation. In a system of linear equations, for example: $$\begin{cases} x - 2y = -7 \\ 3x + y = 0 \end{cases}$$ the solution set consists of the single ordered pair $(-1, 3)$ that satisfies both equations simultaneously. (Number and algebra)
Solve (solution)	A typical **command term** where you need to find the answer to a question using an appropriate method or methods (eg algebraic/graphical/numerical methods).
Solver	A **GDC** function which enables you to solve equations such as exponential equations.
Spanning tree (AHL)	A subset of a graph that connects all the **vertices**. A spanning tree cannot be cyclic and is always a connected graph. Every undirected connected graph has at least one spanning tree. (Geometry and trigonometry)
	(see **minimum spanning tree**)

S

Spatial dimension	The least number of **coordinates** to show any point in a space. (Functions, Geometry and trigonometry)
Spearman's rank correlation coefficient	A statistical test to compare data from two variables that have been put in an order. (Statistics and probability)
Sphere (spherical)	A **three-dimensional** solid which is perfectly round, a ball. (Geometry and trigonometry)
Spiral (AHL)	A **phase portrait** of a **dynamic system** when the **eigenvalues** are complex. You need to be familiar with the spiral, but will not be asked to calculate the exact solutions. For example, the phase portrait of $\frac{dy}{dt} = 2x - 8y, \frac{dx}{dt} = 6x - 7y$ is a spiral. (Calculus)
Spread	How data is distributed and varies from the **mean** – we might describe its **range**, **quartiles**, etc. (Statistics and probability)
Spreadsheet	A computer program which enables you to easily work out many calculations. Used often in financial planning. (Number and algebra)
Square matrix (AHL)	A **matrix** that has equal number of rows and columns. A square matrix has dimensions $n \times n$, where n is the number of rows and columns. For example: $(-3) \quad \begin{pmatrix} 1 & 5 \\ 4 & 1 \end{pmatrix} \quad \begin{pmatrix} x & -2 & y \\ 0 & 0 & z \\ -z & 2y & 2x \end{pmatrix}$ $1 \times 1 \quad\quad 2 \times 2 \quad\quad 3 \times 3$ (Number and algebra)
Square number	The result of multiplying a number by itself. For example, 5 squared is 25 as $5 \times 5 = 25$, so 25 is the square number of 5. (Number and algebra) (see **square root**)
Square residuals (AHL)	The square of the difference between the data values (the points on a scatter diagram) of the **dependent variable** and the curve or **line of best fit**. (Statistics and probability)
Square root	The inverse of a **square number**, signified by $\sqrt{}$, for example $\sqrt{25} = 5$ as $5 \times 5 = 25$ (Number and algebra) (see **square number**)
Stable population (AHL)	When the growth of a population is unchanging, ie the rate of change is zero. In a **phase portrait**, this is seen when there is a stable point. (Calculus)

Standard deviation	A way of measuring the **spread** of data and how the values in a data set vary from the **mean**. The standard deviation is the **square root** of the **variance**. (Statistics and probability)
Standard form	A way of expressing a number as a number between 1 and 10, multiplied by 10 to the power of something. Often used for very large or very small numbers, eg 53000 in standard form would be 5.3×10^4. (Number and algebra) Also called **scientific notation**.
State	Often used as a **command term**. You are just required to write something down without including a calculation or an explanation. For example, when you are asked to state the number of sides of a square, you would write 4.
State (AHL)	The possible outcomes in a **discrete dynamical system**, such as a **Markov chain**. For example, in a Markov chain about sunny and rainy days, the states are Sunny and Rainy. (Statistics and probability)
State space (AHL)	The set of all possible states of a **discrete dynamical system**. (Statistics and probability)
State transition (AHL)	See **transition**. (Statistics and probability)
Stationary point	A point on a curve where the gradient is zero. Signified by a maximum, minimum, or point of inflection. (Calculus) (see **turning point**)
Stationary probability matrix (AHL)	The probability matrix resulting from a steady state – when the probabilities of being in each state of a **Markov chain** approach some limit and are unchanged. (Statistics and probability)
Statistic (of a sample)	One value taken from a **population**. (Statistics and probability)
Statistical inference (AHL)	Making **predictions**, **generalizations**, or **estimates** of a **population** based on information obtained from a **sample**. (Statistics and probability)
Statistics	A branch of Mathematics which works with data values and their analysis. (Statistics and probability)
Steady state probability (AHL)	When the **probability** of being in each state in a **Markov chain** approach some limit and are unchanged from one **transition** to the next. When this occurs, the Markov chain is said to have a **stationary probability matrix**. (Statistics and probability)
Step size (AHL)	The fixed difference in the x-values used to find an approximate solution to first-order differential equations. (Calculus) (see **Euler's method**)
Stochastic process (AHL)	A process in which values are randomly changing over time. A **Markov chain** is an example of a stochastic process. (Statistics and probability)
Stratified sampling (stratified sample)	A method of choosing from a set of data so that the number in each section is represented by a **proportion** of the number in the whole **population**. (Statistics and probability) (see **random sample**)
Strength (of a relationship)	A measure of how closely two variables are related. When looking at **correlation**, we can have **strong correlation**, **weak correlation** and many in-between relationships (neither strong or weak). (Statistics and probability)

Stretch (AHL)	The **transformation** of a graph where a) all x values of the points on a graph are multiplied by the same **scale factor**, p, and the graph enlarges ($p > 1$) or shrinks ($p < 1$) horizontally; or b) all y values of the points on a graph are multiplied by the same scale factor, q, and the graph enlarges ($q > 1$) or shrinks ($q < 1$) vertically. The graph of relationship f under a horizontal stretch with scale factor p The graph of relationship f under a horizontal stretch with scale factor p transforms to the graph of $f(\frac{x}{p})$; the graph of relationship g under a vertical stretch with scale factor q transforms to the graph of $qg(x)$. Using matrices, the geometrical transformation of a horizontal stretch with scale factor p is obtained through matrix $\begin{pmatrix} p & 0 \\ 0 & 1 \end{pmatrix}$; a vertical stretch with scale factor q is obtained through matrix $\begin{pmatrix} 1 & 0 \\ 0 & q \end{pmatrix}$. A matrix $\begin{pmatrix} p & 0 \\ 0 & 1 \end{pmatrix}$ will stretch both horizontally and vertically by the corresponding factors. Multiplying a complex number $z_1 = a_1 + ib_1 = r_1 e^{i\theta_1}$ with another complex number $z_2 = a_2 + ib_2 = r_2 e^{i\theta_2}$, will result to the stretch of z_1 by a factor of r_2 (as well as a rotation by θ_2). (Number and algebra, Functions, Geometry and trigonometry)
Strong (correlation)	When two sets of data which we are comparing are very strongly linked. For example, when working with **Pearson's product-moment correlation coefficient**, strong correlation would be near to 1 or −1. (Statistics and probability) (see **weak (correlation)**)
Strongly-connected graph (AHL)	A graph where there is an **edge** between every possible pair of **vertices**. (Geometry and trigonometry)
Subgraph (AHL)	A graph, usually denoted G', where the **vertices** and **edges** are subsets of another graph, G. (Geometry and trigonometry)
Submatrix (AHL)	The **matrix** remaining after removing one or more rows and/or columns from a matrix A. For example, matrix B below is a submatrix of the original matrix A. $$A = \begin{pmatrix} 32 & 55 & 56 \\ 44 & 12 & 78 \\ 41 & 53 & 89 \end{pmatrix} \quad B = \begin{pmatrix} 12 & 78 \\ 53 & 89 \end{pmatrix}$$ (Number and algebra, Geometry and trigonometry)
Subset	A **set** which is part of another set. For example: {even numbers} is a subset of {whole numbers} (Number and algebra)
Substitute (substitution)	To replace a letter by a number. For example, if you substitute 2 for x and 3 for y in the **expression** $5x + 4y$, you would get $(5 \times 2) + (4 \times 3)$ or 22. (Number and algebra)
Subtraction (subtract)	To take away, minus. For example, 5 subtract 1 is 4. (Number and algebra)
Sufficient condition	Something which must be satisfied for a statement to be true. If the condition is not met, then the statement cannot be true.
Suggest	Often used as a **command term**. You need to give the answer you believe fits best with the evidence you have considered. This might be a solution or a hypothesis, for example.

Sum	1 What you get when you add two numbers or expressions. (Number and algebra)
	2 Another name for a question in Mathematics.
	3 A GDC function.
Sum of square residuals (AHL)	The sum of the squares of the differences between the data values (the points on a scatter diagram) of the dependent variable and the line of best fit (see **square residuals** and **least squares regression**). It is written as SS_{res}. The smaller the sum of square residuals, the better the curve of best fit matches the data. Given points (x_i, y_i) on a scatter graph and a line of best fit, $f(x)$, the sum of square residuals is: $$SS_{res} = \sum_{i=1}^{n}(y_i - f(x_i))^2$$ (Statistics and probability)
Sum to infinity	If a sequence approaches a limit as x approaches infinity, then the limit is the sum to infinity. (Number and algebra)
Surd	A number written using a square root sign which cannot be simplified any further. For example, $\sqrt{5}$ or $\sqrt{3}$. Used to give *exact* answers. (Number and algebra)
Surface area	In a **three-dimensional** shape, the surface area is the area of each of the faces added together, measured in square units. (Geometry and trigonometry)
Survey (AHL)	The process of collecting data from a given group of people or things. Surveys can use **questionnaires**, interviews, observations or experiments to collect data. For example, a survey could be a questionnaire about people's feelings towards a new government policy or a collection of data about the concentration of a chemical in a water source. (Statistics and probability)
Symbol (symbolically)	A character or sign used instead of words. In Mathematics, the use of mathematical notation such as +, ÷ , etc.
Symmetrical (symmetry)	When a shape is divided into two equal parts, which are mirror images of each other.
System of equations	When you have more than one equation containing common **variables** to be solved simultaneously. Equations in the system can be **linear** or **non-linear**. (Number and algebra)
System of linear equations	A **system of equations** but not involving anything which would not form a line. (Number and algebra)
Systematic sampling (sample)	A method in statistics of choosing from a population where the starting point is random but where subsequent members of the population are selected with a fixed, periodic interval. (Statistics and probability)

T

Table of least distances (AHL)	Given a table of distances between cities in the **travelling salesman problem**, you can use **Prim's algorithm** on the table to find the routes with the least distances. (Geometry and trigonometry)
Table of outcomes	The results of an experiment presented in a tabular form. (Statistics and probability)
Tangent	A straight line outside a circle or curve which touches the curve at one point only. (Geometry and trigonometry, Calculus)
Tangent (tan) function	A trigonometric function. The tangent of an angle can take any value and a graph can be drawn to represent it. (Functions) (see **cosine (cos) function**, **sine (sin) function**)
Tangent (tan) ratio	The length of the **opposite** side divided by the **adjacent** side in a right-angled triangle. (Geometry and trigonometry)
t-distribution (AHL)	An approximation of the **normal distribution** (it is very similar shaped) used for small samples ($n < 30$) when the population variance is unknown. The value of the test statistic, t, is found using the formula $t = \frac{\bar{x} - \mu}{s_{n-1}/\sqrt{n}}$, which is given in the formula book. You should be able to use the t-distribution to find the **probability** that the value of a random variable, T, lies in a certain range of values, $P(T \leq t)$, and also to find the boundary value, t, if given the cumulative probability. (Statistics and probability)
Term	1 A combination of numbers and variables, such as $5x^2$. There may be several terms in an **expression** separated by + or – signs. 2 A value in a **sequence**. (Number and algebra)
Term to term rule	How to get from one value in a **sequence** to another. For example, for odd numbers, the term to term rule would be +2. (Number and algebra)
Terminal point (AHL)	The point at which a vector finishes. (Geometry and trigonometry) (see **initial point**)
Terminology	The group of words used in a particular subject to explain and express methods, etc. In Mathematics, it's important for you to use appropriate terminology in classroom discussions and in your written work, particularly your **IA**, to achieve the highest marks.
Test statistic	A statistic that is taken from a **sample**, and used in a **hypothesis test**. In the hypothesis test, it is the statistic used to compare the data from the sample with the probability of that outcome happening by chance. For example, in a normal distribution, the test statistic is the Z-score. In a **t-distribution**, the test statistic is the t-score. (Statistics and probability)
Test-retest (AHL)	When the same test is given twice to the same people at different times to determine the **reliability** of the results. The reliability is determined by the strength of the **correlation** between the results of the test and the retest. (Statistics and probability)
Theorem	A result or assertion which someone has proved to be true, eg **Pythagoras' theorem**.
Theoretical probability	How you expect an experiment to turn out; the likelihood or **probability** based on reasoning. Written as a **ratio**, we say the theoretical probability of throwing a 4 on a **die** is 1/6. This would probably not happen in an experiment. (Statistics and probability) (see **experimental probability**, **empirical probability**)

Theta	A letter of the Greek alphabet which is often used to represent an angle. Written as θ. (Functions, Geometry and trigonometry)
Thinker	The IB learner profile suggests ten qualities that students should be aiming for. One of these is to be a 'thinker' which can involve being able to use your mind to work out situations and make decisions. In your IB Mathematics studies, this can mean working out how a real-life problem is related to Mathematics.
Three-dimensional (3D)	Shapes like **cubes**, **cuboids**, **spheres**, etc are three-dimensional as they are not flat shapes; they have length, breadth and height. (Geometry and trigonometry)
Three-dimensional shape (3D shape)	A shape like a **cube**, **cuboid**, or **sphere** which is not flat, and which can have length, breadth and height. (Geometry and trigonometry)
Three-figure bearing	An angle which signifies the position of one point in relation to another. It is measured from a North line and always measured clockwise. If it is less than 100, a 0 is put in front, so that the bearing always has 3 figures, eg 088°. (Geometry and trigonometry)
Time-shift (AHL)	To transform (or change) an equation of motion that is dependent on time to account for a different starting time. Given a function $f(t)$, $f(t-a)$ is a time-shift of a units where a is a unit of time, eg seconds. (Geometry and trigonometry)
Toxic waste dump problem	A problem which is used to illustrate the use of **Voronoi diagrams**. Given the coordinates for a number of cities, you can use a Voronoi diagram to find a point furthest away from all the cities which would be the best location for a toxic waste dump. (Also called large empty circle problem.) (Geometry and trigonometry)
Trace	A **GDC** function. It enables you to find a y value of a graph, when you input the x value.
Trail (AHL)	A route taken from one **vertex** to another (a **walk**) that does not revisit any **edges**. It is a sequence of edges and **adjacent vertices**. (Geometry and trigonometry)
Trajectory (plural trajectories) (AHL)	The path that the solution to a **dynamic system** follows in a **phase portrait**. The trajectory represents the way in which the solution changes with respect to time. Given an initial state, the trajectory is the set of points that represent the future states. (Calculus)
Transformation	A way of moving a shape – this can be a **reflection**, **rotation**, **translation**, or **enlargement**. (Geometry and trigonometry, Functions, Statistics and probability)
Transient state (AHL)	A state in a **Markov chain** where the **probability** of returning to that state is a non-zero probability less than one. See **recurrent state** for an example. (Statistics and probability)
Transition (AHL)	The movement between states (or a state returning to itself) in a **Markov chain**. For example, predicting the weather can be seen as a Markov chain where it will either be a cloudy day or a sunny day the next day. If it is a sunny day today, it can transition to be cloudy tomorrow or to be sunny again tomorrow. (Statistics and probability)
Transition diagram (AHL)	A diagram that shows the possible **transitions** (or changes) between states in a Markov chain and the associated **probability**. (For an example, see **Markov chain**.) The transitions and associated probabilities can also be shown using a **transition matrix**. (Statistics and probability)

A-Z for Maths: Applications and interpretation (published by Elemi)

T

Transition matrix (AHL)

1. In a **Markov chain**, a matrix that shows the transitions between states and the associated probabilities. For example, given states A and B, the transition matrix $\begin{pmatrix} P_{A,A} & P_{A,B} \\ P_{B,A} & P_{B,B} \end{pmatrix}$ shows the possible transitions between states and the probabilities of those transitions where $P_{A,A}$ is the probability of A returning to A, $P_{A,B}$ is the probability of moving from A to B and so on. (Statistics and probability)

2. In **graph theory**, a matrix showing the probabilities of each step in a random **walk**. The probability of moving between vertices i and j is given in the (i,j)th entry in the matrix. You should be able to construct a transition matrix for a strongly-connected, directed or undirected graph. (Geometry and trigonometry)

Transition probability (AHL)

The **probability** of a **transition** (or a change in state) in a **Markov chain**. (Statistics and probability)

Translation

When a shape moves but does not change shape or **orientation** – it slides. (Functions, Geometry and trigonometry)

Translation vector

A way of explaining how a shape moves, signified by two numbers in brackets on top of one another, where the top number represents a move to the right (or left, if negative) and the bottom number a move up (or down, if negative). (Functions, Geometry and trigonometry)

Trapezoid

A **two-dimensional quadrilateral** with one pair of parallel sides. (Called trapezium (*plural* trapezia) in the UK.) (Geometry and trigonometry, Calculus)

Trapezoidal rule (*or* trapezoid rule)

A way of finding the area under a graph by dividing the area into trapezia. It allows you to find an approximate solution to a **definite integral**. (Calculus)

Travelling salesman problem (AHL)

A problem that supposes a travelling salesman is trying to find the shortest possible total route between a number of cities that returns to the original city. It is a typical problem that is approached using graph theory, and is an example of finding a **Hamiltonian cycle of least weight**. It is not possible to find an exact solution; instead you find the smallest and largest values the actual solution could take. The methods used to solve the problem can be applied to other contexts, such as computer networks. You should be able to model a practical problem as a travelling salesman problem and use the same methods to reach a solution. (Geometry and trigonometry)

Traversable (graph) (AHL)

A graph that has a **Eulerian trail**, so it has a path that visits every edge exactly once. You can visit the same vertex more than once. You can trace a traversable graph without lifting your pencil. (Geometry and trigonometry)

Tree (AHL)

An **undirected graph** where any two vertices are connected by only one **path**. A tree is always a connected acyclic graph. (Geometry and trigonometry)

Tree algorithm (AHL)

An **algorithm** used to find a **tree** in a graph, such as **Prim's** and **Kruskal's algorithms**. (Geometry and trigonometry)

Tree diagram

A visual representation (shaped a little like a tree) that is used to help calculate the probability of something happening. (Statistics and probability)

Trial

A single experiment in **probability**. (Statistics and probability)

Triangle law (AHL)

A method of adding **vectors**. If you are given the **magnitude** and direction of two vectors, you can use trigonometry to find the magnitude and direction of the **resultant** vector. For example, given the vectors a and b, you can find the length of the resultant vector, c using the **cosine rule** and the direction of c using the **sine rule**. (Geometry and trigonometry)

Triangle sum theorem	The sum of the three angles of any triangle is 180 degrees. (Geometry and trigonometry)
Triangulation	A method of using angles you do know to calculate angles and sides that you do not know, when working with triangles. Used in real-life situations like surveying and navigation. (Geometry and trigonometry)
Trigonometric expression	A mathematical **expression** that contains **trigonometric functions**, such as $2 \sin 3x + 5 \tan y$. (Geometry and trigonometry)
Trigonometric function	1. A function that describes the relationship between the angle of a **right-angled triangle** and the ratio of the length of two of its sides. 2. A function that involves the **trigonometric ratios** (sine, cosine and/or tangent) such as $f(x) = 2 \sin 3x$. (Functions, Geometry and trigonometry) (see **sine function**, **cosine function** and **tangent function**)
Trigonometric ratio	Relates the angle of a **right-angled triangle** to the ratio of the length of two of its sides. (Geometry and trigonometry) (see **sine ratio**, **cosine ratio** and **tangent ratio**)
Trigonometry (trigonometric)	The study of triangles which involves methods of finding the angles and the lengths of sides of triangles. (Geometry and trigonometry)
Trillion	A million millions (ie one with 12 zeroes). (Number and algebra)
Trillionth	A million millionth. (Number and algebra)
Trivial	Something that is very simple and is not interesting. Often a zero solution (ie $x = 0$) is considered trivial. (see **non-trivial**)
t-test	A statistical method to compare sets of data, most commonly used where there is a **normal distribution**. (Statistics and probability) (see **hypothesis test**)
Turning point	A point where the **gradient** is zero, having gone from negative to positive gradient or vice versa. Shown on a graph as a maximum or minimum. (Calculus)
Two-dimensional (2D)	Shapes like squares, rectangles, etc are two dimensional as they are flat shapes. We only see two dimensions (eg height and width, but no depth). (Geometry and trigonometry)
Two-tailed test	Used in **hypothesis testing** (eg as part of a *t*-test) where you are interested in the results at both ends or 'tails' of the distribution of data. You check the calculated result against two values, to see if it is greater than one or lower than the other. (Statistics and probability) (see **one-tailed test**)
Type I error (AHL)	Rejecting the **null hypothesis** in a **hypothesis test** when in fact the null hypothesis is true. (Statistics and probability)
Type II error (AHL)	Not rejecting the **null hypothesis** in a **hypothesis test** when in fact the null hypothesis is false. (Statistics and probability)

U

Unbiased (AHL) — Showing no opinion on something. For example, to carry out an unbiased **survey**, you need to sample people from different age ranges, locations, etc. (Statistics and probability)

(see **unbiased estimate**)

Unbiased estimate (AHL) — When the value of a statistic calculated from a sample is equal to the value of the same statistic from the population. (Statistics and probability)

(see **biased estimate**)

Unbiased estimator (AHL) — A statistic from a **sample** that gives an **unbiased estimate** – ie one in which the statistic from a sample is equal to that same statistic from the **population**. (Statistics and probability)

Unbounded — Having no limits; not **bounded**. (Geometry and trigonometry)

Uncertainty — Where one cannot be sure of the truth or the exact value. For example, the accuracy of a measurement depends on the person measuring and can be uncertain. (Number and algebra)

Undefined (AHL) — An **expression** or **operation** which does not have meaning and so cannot be reasonably interpreted.

In the set of real numbers, the following are undefined:
- division by zero
- the square root of a negative real number
- the logarithm of a negative number.

A real **function** $f(x)$ is undefined for all values of x out of the **domain**. So functions such as rational functions, logarithmic functions and those including **radicals** will not be defined on the full set of real numbers. (Number and algebra, Functions, Calculus)

(see **domain restriction**)

Underestimate — When the amount you roughly calculate is less than the real value. (Number and algebra, Functions, Calculus)

Undirected graph (AHL) — A graph where the **edges** are not given a direction. (Geometry and trigonometry)

Uniform average rate (AHL) — When events occur in a particular region or time period at a constant rate. This is an important aspect of a **Poisson distribution**. (Statistics and probability)

Union — Used with **sets** – the union of two sets is all elements that are in one, or the other, or both. (Remember if elements are repeated in both sets, they must only be written once in a **Venn diagram** or in the set that is the union of the two sets.) (Number and algebra)

Unique (solution) — If a problem has one and only one possible answer, it has a unique solution. (Number and algebra)

Unit —
1. A general term to refer to 'one' or 'ones'; the digit just before the decimal point in a number. (Number and algebra)
2. A standard quantity that measurements are written in, eg cm, m². (Geometry and trigonometry)

Unit circle (AHL) — A circle with a unit radius. It is usually a circle with radius of 1 centred at the origin (0, 0). It can be used to define **sine**, **cosine** and **tangent**. (Geometry and trigonometry)

Unit normal vector (AHL) — A **vector** with a **magnitude** of 1 that is perpendicular to a given plane. (Geometry and trigonometry)

Unit vector (AHL) — (see **direction vector**) (Geometry and trigonometry)

Universal set	In **sets**, the group from which all members of any set are taken. In a **Venn diagram**, where the sets are represented by circles, the universal set is a rectangle surrounding the circles. (Number and algebra)
Unpaired samples	When the data given is unrelated. (Statistics and probability)
Unweighted graph (AHL)	A graph that is not weighted, ie the **edges** are given no particular value or weight. (Geometry and trigonometry)
Upper bound	1 The largest number possible. For example, when looking at 500 to the nearest 100, the upper bound is given as 550, even though it is really 549.99999… (Number and algebra)
	2 The largest value a measurement can take. For example, an angle in a triangle must be less than 180°. (Geometry and trigonometry)
	(see **lower bound**)
Upper boundary	Calculation of halfway between the top of one **class** and the bottom of the class above it. For example, for 5–8 and 9–12 the upper boundary of the 5–8 class is 8.5. (Statistics and probability)
	(see **lower boundary**)
Upper quartile	Where a data set is divided up into quarters (or **quartiles**). The upper quartile is the number which is three quarters of the way up from the lowest point in the data. This can either be calculated from a list of numbers or shown on a **cumulative frequency graph**. (Statistics and probability)
	(see **lower quartile**)
Upper tail test	Testing whether the test statistic is greater than a certain value in an **hypothesis test**, eg in a **chi-squared test for independence**. (Statistics and probability)

V

Valence		The number of edges at a vertex of a **Voronoi diagram**. (Geometry and trigonometry)
Valid (validity)		True and without **bias**. At AHL, you would be expected to know the difference between validity and **reliability**. (Number and algebra, Statistics and probability)
Validity test (AHL)		A measure of how well the data from a **sample** corresponds to data in a **population**. Validity can be tested for **content validity** and for **criterion-related validity**. (Statistics and probability)
Value	1	In Mathematics, an amount that has been worked out, expressed as a number.
	2	In more general terms, the amount an item is worth; how much it costs or would sell for if sold.
Variable		An amount represented by a letter such as x or y, that can take several **values**. Even if it only has one value, it is still called a variable. (Number and algebra)

Variable matrix (AHL)

A matrix in which all **entries** are **variables**, often used when writing **systems of linear equations** in matrix form. In this case, the variable matrix is the matrix formed by the variables of the system of linear equations. A system of two linear equations with unknown variables x and y is written as:

$$\begin{cases} a_{11}x + a_{12}y = b_{11} \\ a_{21}x + a_{22}y = b_{21} \end{cases}$$

where $a_{11}, a_{12}, a_{21}, a_{22}, b_{11}, b_{21}$ are real numbers. The system is written in matrix form as:

$$\begin{pmatrix} a_{11} & a_{12} \\ a_{21} & a_{22} \end{pmatrix} \begin{pmatrix} x \\ y \end{pmatrix} = \begin{pmatrix} b_{11} \\ b_{21} \end{pmatrix}$$

or

$$AX = B$$

where

$$X = \begin{pmatrix} x \\ y \end{pmatrix}$$

is the variable matrix. (Number and algebra)

(see **answers matrix**, **coefficient matrix**, **constant matrix**)

Variance	A measure of how spread out numbers are in relation to the **mean**. It is the **square** of the **standard deviation**. (Statistics and probability)
Variation	A change in one **value**, related to a change in another. Variation can be described as **direct proportion** (or direction variation) or **inverse proportion** (or inverse variation). (Statistics and probability)
Vector	A way of showing both distance and direction of a line. It can be signified by two numbers in brackets on top of one another, where the top number represents a move to the right (or left, if negative) and the bottom number represents a move up (or down, if negative). (Geometry and trigonometry)
Vector algebra (AHL)	The operations of vector addition, **scalar multiplication**, **scalar product** and **vector product**. (Geometry and trigonometry)
Vector equation of a line (AHL)	Given a point travelling in a straight line, the vector equation of the line (also called the position vector) is $\boldsymbol{r} = \boldsymbol{a} + \lambda \boldsymbol{b}$, where $\boldsymbol{a}$ is the starting position, λ is a scalar, and $\boldsymbol{b}$ is the direction vector. (Geometry and trigonometry)
Vector field (AHL)	(see **slope field**) (Calculus)

Term	Definition
Vector product (AHL)	An operation on two **vectors** that gives a vector that is perpendicular to both vectors. It is denoted with the symbol × and is said as 'a cross b'. It is also called the cross product. $a \times b = \|a\|\|b\| \sin \theta \, n$ where θ (is the angle between a and b and n is the **unit normal vector** for the plane containing a and b. Use the **right-hand screw rule** to determine the direction of n. The **magnitude** of the vector product is the area of a parallelogram bounded by the vectors a and b. Note that $a \times b$ does not equal $b \times a$. (Geometry and trigonometry)
Vector theory (AHL)	Using **vectors** in **mathematical modelling**, for example, to model the movement of a **projectile**. (Geometry and trigonometry)
Velocity (AHL)	The rate of change of the position of an object per unit of time. Velocity is a vector quantity; it has magnitude (which we call speed) and direction. (Geometry and trigonometry, Calculus)
Velocity function (AHL)	A function that describes the **velocity** of an object x at time t. The velocity function is the **first derivative** with respect to time of the **displacement function**. If $x = x(t)$ is the displacement function, then $v(t) = \frac{dx}{dt}$ is the velocity function. (Calculus)
Venn diagram	A visual representation used in **sets** to show in which set elements belong and their relationship. A Venn diagram consists of a rectangle (see **universal set**) surrounding circles which represent the sets. (Statistics and probability)
Verify (verification)	To prove something is true through evidence. Often used as a **command term** where you need to provide evidence to prove that a result is correct.
Vertex (*plural* vertices)	1. On a graph, the vertex is a maximum or minimum point. (Functions) 2. Where two lines or edges join in a **2D** or **3D** shape. (Geometry and trigonometry) 3. On a **Voronoi diagram**, where the boundaries of three or more Voronoi cells meet. (Geometry and trigonometry) 4. (AHL) In **graph theory**, a point in a graph. (Geometry and trigonometry)
Vertex disjoint (AHL)	Two **paths** in a graph that do not share a **vertex**. (Geometry and trigonometry)
Vertex set (AHL)	The set of **vertices** in a graph. (Geometry and trigonometry)
Vertical	Going from north to south or top to bottom. A vertical line is perpendicular to a **horizontal** line. (Functions, Geometry and trigonometry)
Vertical asymptote	A line (often drawn as dotted) which a graph approaches but never reaches. For example, the graph of $y = 1/(x - 2)$ has a vertical asymptote of $x = 2$. (Functions)
Vertical line test	A test to prove that a graph is a **function**. If you draw a line through the graph from top to bottom and it cuts the graph more than once, then it is not a function. (Functions)
Vertical stretch	When a function $f(x)$ is transformed into $af(x)$ the graph undergoes a vertical stretch and all of its y coordinates are multiplied by a. For example, in the stretch $f(x)$ to $2f(x)$ the y coordinates are multiplied by 2. (Functions) (see **horizontal stretch**)
Vertices	The plural form of **vertex**. (Functions, Geometry and trigonometry)
Visualize (visualization)	To make a problem more concrete in your mind by imagining or seeing how it looks in diagrammatical form.

V

Volume	The amount of space inside a **3D** shape. This is measured in cm³, m³, litres, etc. (Geometry and trigonometry)
Volume of revolution (AHL)	The volume contained within a curve when it is rotated along the axes between two values. $V = \int_a^b \pi y^2 dx$ gives the volume of revolution around the **x-axis**. $V = \int_a^b \pi x^2 dy$ gives the volume of revolution around the **y-axis**. (The formulae are given in the formula booklet so you do not need to memorize them.) (Calculus)
Voronoi cell	(see **cell (on a Voronoi diagram)**) (Geometry and trigonometry)
Voronoi diagram	A visual **2D** representation that shows regions (called cells) which comprise all points that are nearer to a **site** in the region than any other specified site of the diagram. (Geometry and trigonometry)
Voronoi face	(see **face (on a Voronoi diagram)**) (Geometry and trigonometry)
Voronoi region	(see **cell (on a Voronoi diagram)**) (Geometry and trigonometry)
Voronoi tessellation	Another term for a **Voronoi diagram**. (Geometry and trigonometry)
Voronoi vertex	Where the boundaries of three or more Voronoi cells meet. (Geometry and trigonometry)

Walk (AHL)	A route taken from one **vertex** to another. It is a sequence of **edges** and **adjacent vertices**. A walk can revisit edges and vertices. (Geometry and trigonometry)
Weak (correlation)	When two values or sets of data which we are comparing do not have a strong link. For example, with **Pearson's product-moment correlation coefficient**, weak correlation would be indicated by a result close to 0. (Statistics and probability) (see **strong (correlation)**)
Weight	How heavy something is – usually measured in kg or tonnes. (Geometry and trigonometry)
Weighted adjacency tables (AHL)	A table used to represent the connections between **vertices** of a **weighted graph**. It is similar to an **adjacency matrix** but where the vertices are connected, the values are the weights of the edges rather than 1s. (Geometry and trigonometry)
Weighted complete graph (AHL)	A graph where each pair of **vertices** is connected by an **edge** and the edges are given a value or weight. (Geometry and trigonometry)
Weighted graph (AHL)	A graph in which the **edges** are given a value or weight. For example, in the **travelling salesman problem**, the edges are assigned a weight relative to their distance. (Geometry and trigonometry)
Wheel graph (AHL)	A graph where every **vertex** connects to one particular vertex. The graph looks like a wheel with a central vertex and the connecting edges look like spokes (of a wheel). (Geometry and trigonometry)
Window	On a **GDC**, the window is where you can put in ranges of values of x and y so that you can choose how your graph appears.
With respect to	Used in calculus to express **differentiation** in relation to the **variable**. So dy/dx is another way of saying y is differentiated with respect to y; and ds/dt is s differentiated with respect to t. (Calculus)
Write down	Often used as a **command term** in your IB studies. You write your answer, usually by finding it in the information you are given which would not usually require much calculation, if any. You would not normally need to show any workings either.

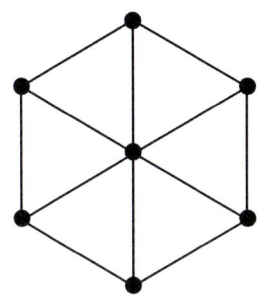

X

x-axis — On a graph, the x-axis is a **horizontal** line which goes across the graph and is referred to as $y = 0$. (Geometry and trigonometry)

x-intercept — The point where a graph cuts the x-axis. (Functions)

Y

Yates' continuity correction — When doing a **chi-squared test for independence**, this is used to correct an error (an over-estimation) of the test when the degree of freedom = 1. (Statistics and probability)

y-axis — On a graph, the y-axis is a **vertical** line which goes up the graph and is called $x = 0$. (Geometry and trigonometry)

y-intercept — The point where a graph cuts the y-axis. (Functions)

Z

Zero (of a function) (*plural* zeroes) — The value(s) of x which make a function zero. For example, for $x^2 - 7x + 10 = 0$, the zeroes are 2 and 5. (Number and algebra, Functions)

Zero matrix (AHL) — A matrix in which all the **elements** are zeros ($a_{ij} = 0$). Unlike the **identity matrix**, which is always a **square matrix**, a zero matrix can have any **dimension**. For example:

$$\begin{pmatrix} 0 & 0 \\ 0 & 0 \end{pmatrix} \quad \begin{pmatrix} 0 \\ 0 \\ 0 \end{pmatrix} \quad (0 \quad 0)$$

The symbol O is often used to represent a zero matrix. The role of the zero matrix in matrix addition and multiplication is the same as the role of 0 in the real numbers set. In other words:

$$A + O = O + A = A$$

and:

$$AO = OA = O$$

(Number and algebra)

Zero vector (AHL) — A **vector** with a **magnitude** of zero. Also called a null vector. (Geometry and trigonometry)

Zoom — On a **GDC**, the zoom button enables you to move the view of a graph closer or further away.